# *The Destiny of the Universe*

## In Pursuit of the Great Unknown

# *The Destiny of the Universe*

## In Pursuit of the Great Unknown

Dr. Gerard M. Verschuuren

Paragon House
St. Paul

First Edition 2014

Published in the United States by
Paragon House
1925 Oakcrest Avenue, Suite 7
St. Paul, MN 55113

www.ParagonHouse.com



Library of Congress Cataloging-in-Publication Data

Verschuuren, G. M. N. (Geert M. N.)
The destiny of the universe : in pursuit of the great unknown / by Gerard M. Verschuuren. -- First Edition.
pages cm
Includes index.
Summary: "Using the latest scientific discoveries and theories of astronomy and genetics, the author discusses the nature of life and the universe and asks big questions like, 'Why are we here?' It is both a critique of scientific atheism, and a scientific argument for the "unknown" that people call 'God'"-- Provided by publisher.
ISBN 978-1-55778-908-2 (pbk. : alk. paper) 1. Metaphysics. I. Title.
BD111.V47 2014
110--dc23
2013044967

Manufactured in the United States of America

10 9 8 7 6 5 4 3 2 1

The paper used in this publication meets the minimum requirements of American National Standard for Information Sciences—Permanence of Paper for Printed Library Materials, ANSIZ39.48-1984.

*Dedicated to my wife, Trudy,*
*whose unwavering and loving support*
*has played an important role in my growth process.*

# Acknowledgments

I wish to express my gratitude to those nameless who steered and corrected me in the process of preparation for this book. In particular, I want to mention Francisco Ayala (University of California at Irvine), Stephen Barr (University of Delaware), the late Joseph Flanagan (Boston College), Owen Gingerich (Harvard University), Peter Kreeft (Boston College), and also some great minds from my past in the Netherlands who shaped my mind before they passed away: the philosopher Cornelis van Peursen and the biologist Marius Jeuken (both of Leiden University), and the human biologist John Huizinga, M.D. (of Utrecht University Medical School). And last but not least I want to thank our daughter Elizabeth Frangiosa, Attorney at Law, who helped me in polishing the text.

Obviously, they are not responsible for the outcome; if I erred, it's entirely my doing. They and many others make me realize that originality often consists in the capacity of forgetting about your sources.

# Contents

# Preface

This book is not about astronomy, not even about science per se, but about the Great Unknown beyond and behind all that we can see through our telescopes and microscopes. Although there is a lot of science in this book, albeit on a simplified level, it is mainly a critical and intellectual philosophical journey, starting in the world of science, but ultimately in pursuit of the Great Unknown that has become more and more known in the lives of so many people.

The Great Question is this: Is there someone out there to provide for us? It has been a life-long process for me to finally say wholeheartedly yes to this question. Let me be honest with you and place my cards open on the table. I am a practicing Catholic; if that makes me suspect in your eyes, please give me a chance, or at least the benefit of the doubt. I am not speaking in this book to those inside the fold of any church, but to those living in its diaspora, where I once was their neighbor.

It is a book for those living in the barren interstellar space of our universe, surrounded by black holes, quasars, and pulsars, and feeling quite lost in the vastness of this universe. Since I have been there myself, I know it is a place of extreme coldness—3° Kelvin, -270° Celsius, or -454° Fahrenheit. It is the cradle of skepticism, relativism, secularism, nihilism, and atheism—sometimes under a thin veneer of religion. It is for those people that I wrote this book, for all those doubters, skeptics, and even nihilists in our midst, as long as they do not have a closed mindset yet.

How did I find my way back to our warm and people-friendly planet? Fortunately I had some memories left from my stay on this planet. Those memories gave me enough incentives to begin my journey. Let me invite you and take you on my journey back to our Judeo-Christian heritage. If you stop halfway, I will understand. You may be ready to resume this journey once you reach a next stage in your life, in the same way as I had to get ready for each next step. Take this book as an autobiography—possibly as a reflection of your own autobiography as well.

For years, I let my surroundings dictate what my worldview should be—nurtured by science—until I was able to notice its shortcomings. So I began to question its merits in order to replace my narrow-minded worldview with a much richer, more all-comprehensive, and more realistic view of the world. It is my strong conviction that it is not so much science but a misguided philosophy that caused me to go astray, so perhaps a better kind of philosophy can bring us back on the right path again.[1] As for me, it was not that I started ignorant, but I knew so much that is not so, until philosophy, and finally religion, brought me closer to the Truth. It is my hope that, as one grows older, one's outlook tends to become broader and richer.

During my training as a scientist, I was hardly ever taught to look beyond the boundaries of science. In this day and age, studying a specific science takes so much time that almost no time is left for other issues; besides, those teaching science usually want to stress its strengths, not its weaknesses. I had to discover the limitations of science myself—I am sure with God's help.

An added problem is that science cannot be studied by scientists from within. In order for us to study science—its internal structure, its foundations, its assumptions, as well as its limitations—we need to adopt a bird's-eye view, a so-called meta-level, a philosophical perspective, a metaphysical viewpoint—a science of science, if you will. Unfortunately, it seems many scientists have never reached such a "high altitude." They tend to stare at that square inch, nanometer, or micron they are working on and feel comfortable with, while forgetting that there is so much more beyond their limited scope. They work so hard that they have hardly any time left for serious thinking.[2]

The order of chapters in this book reflects more or less the course of my life. Each new chapter represents another stage in what I went through. It was a path starting in the desert of science but bringing me closer and closer to the oasis of God. Perhaps you will recognize some of these stages in your own life. It may be a journey of intellectual struggling, but with a good ending. I tried not to be too much of a scientist for philosophers, and not too much of a philosopher for scientists. I deeply believe they can learn

1. In his autobiography, C.S. Lewis writes something similar: "You will understand that my [atheism] was inevitably based on what I believed to be the findings of the sciences." In *Surprised by Joy: The Shape of My Early Life* (Houghton Mifflin Harcourt, 1995), 97.

2. I borrowed this statement from the Nobel Laureate and biophysicist Francis Crick: *What Mad Pursuit: A Personal View of Scientific Discovery* (New York: Basics Books, 1990).

from each other.

In the first chapter ("Living in the Midst of Black Holes"), I am going to sketch the seemingly solid structure of the scientific framework that most of us have been brought up with. It certainly appears to be a solid structure—until the waves of life start to shake it.

In the second chapter ("An Opening in the Fence"), it begins to dawn on us that the world of science is not as closed as it appears, and that human beings must be more than what science tells us that they are. That is ultimately the beginning of a deeper reflection on each one of us.

In the third chapter ("Are We Alone?"), we may get a vague suspicion that there is more "out there" that science has no access to, but nevertheless an inescapable reality. Was our universe perhaps *created*, or is that just a shaky belief? The leap of faith is not a plunge into the irrational, and certainly not a plunge into the unreasonable, leaving the rest of us behind. That is why we need some sound philosophical reflection here.

In the fourth chapter ("Cues from Beyond"), we are still in a skeptical stage, but with a hint of wonder: Is there someone beyond and behind our telescopes and microscopes? It is a search for something absolute in a world of relativity, for something immaterial in a world of mere matter, for an oasis in a seemingly God-forsaken desert. The search for the Great Unknown is definitely on.

In the fifth chapter ("Knowing the Great Unknown"), we reluctantly acknowledge, but now with good reason, that there may be some kind of providence in this universe: all-present, all-knowing, and all-powerful. What would that providence be like and how would such a thing be possible in our seemingly closed universe? Can the physical universe ever become a meaningful universe?

In the sixth chapter ("All-Loving Providence"), it is time to become a bit more theological without ever leaving the sound and strong reasoning power of philosophy behind us. I hope by then I have not lost you, for then you would miss out on some live-saving answers to some life-size questions.

# 1. Living in the Midst of Black Holes

Science gives, and science takes away, at the same time. The more we know about the universe, the less we seem to know about ourselves. We seem to almost be drowning in the vastness of the universe. The late American astrophysicist Carl Sagan is known to have called our planet "a speck of dust in the universe"[1]—"all that is, or was, or ever will be."[2] It is this speck of dust that we call home. If this planet is a speck of dust, imagine us living on it—we are even tinier specks of dust. We become tinier by the minute and end up as star dust. No wonder we seem to be living in a God-forsaken universe.

The idea is basically very old—it is the ancient pagan philosophy of materialism. It holds that the only thing that exists is matter; that all things are composed of matter and dust, the mere result of material interactions. In other words, matter is supposedly the only substance in this universe; and thus physical matter has become the only fundamental reality.

In antiquity, the atomic theory of Greek philosophers had this notorious aura of claiming, not only that atoms are everywhere, but also that they are all there is. That is how materialism started, and it has never really left us since. For some enigmatic reason, materialism has quite a spiritual appeal to it. However, the curious thing about it is that, if you deny the existence of anything immaterial, you also deny the existence of your very own denial, for denials are certainly immaterial as well. But let us not get ahead of ourselves.

1. Carl Sagan, *Pale Blue Dot: A Vision of the Human Future in Space* (New York: Ballantine Books, 1997).
2. Carl Sagan, *Cosmos, One Voice in the Cosmic Fugue* (New York: Random House, 2002).

## Living in a Closed World

Here is what science tells us about the world we live in. We are surrounded by black holes,[3] quasars,[4] and pulsars.[5] Where do they come from? Nowadays we know that our universe most likely started with the Big Bang, almost fourteen billion years ago. In 1929, Edwin Hubble discovered that the distances to far away galaxies were generally proportional to their red shifts[6]—an idea originally suggested in 1927 by the Belgian priest, astronomer, and physicist George Lemaître of the Catholic University of Louvain.[7]

Hubble's observation was taken to indicate that all very distant galaxies and clusters have an apparent velocity directly away from our vantage point: the farther away, the higher their apparent velocity. In 1931, Lemaître went even further and suggested that the evident expansion of the universe, if projected back in time, meant that the further in the past the smaller the universe was, until at some finite time in the past all the mass of the universe was concentrated into a single point, a "primeval atom" where and when the fabric of time and space came into existence.[8]

The English astronomer and mathematician Fred Hoyle is credited with coining the term Big Bang during a 1949 radio broadcast. Currently, the Big Bang theory is the prevailing cosmological model that explains the early development of the universe. According to this theory, the universe was once in an extremely hot and dense state which expanded rapidly. This rapid expansion caused the universe to cool and resulted in its present

3. A black hole is a region of space-time where gravity prevents anything, including light, from escaping.

4. A quasar is a compact region in the center of a massive galaxy surrounding its central super-massive black hole. A Nobel Prize for the study on quasars was given to the British radio astronomers Anthony Hewish and Martin Ryle, in 1974.

5. A pulsar is a highly magnetized, rotating neutron star that emits a beam of electromagnetic radiation. A Nobel Prize for the study on pulsars was given to the American astrophysicists Joseph H. Taylor and Russell A. Hulse, in 1993.

6. Red shift happens when light seen coming from an object that is moving away is proportionally increased in wavelength, or shifted to the red end of the spectrum.

7. G. Lemaître, "Un Univers homogène...," *Annales de la Société Scientifique de Bruxelles* (April 1927), 47: 49.

8. G. Lemaître, "The Beginning of the World from the Point of View of Quantum Theory," *Nature* 127 (1931), 3210, 706.

continuously-expanding state. Once it had cooled sufficiently, its energy was allowed to be converted into various subatomic particles, including protons, neutrons, and electrons. Giant clouds of these primordial elements would then coalesce through gravity to form stars and galaxies, and the heavier elements would be synthesized either within stars or during supernovae.[9] Interestingly enough, the 92 elements we find on earth can be found all over the universe.

A combination of observations and theory suggests that the first quasars and galaxies were formed about a billion years after the Big Bang, and since then larger structures have been forming, such as galaxy clusters and super-clusters. Populations of stars have been aging and evolving, so that distant galaxies (which are observed as they were in the early universe) appear very different from nearby galaxies (observed in a more recent state). Moreover, galaxies formed relatively recently appear markedly different from galaxies formed at similar distances but shortly after the Big Bang.

After the discovery of the cosmic microwave background radiation in 1964, and especially when its spectrum (i.e., the amount of radiation measured at each wavelength) was found to match that of thermal radiation from a black body,[10] most scientists had become fairly convinced that some version of the Big Bang scenario best fits observations. The Big Bang theory depends on at least one major assumption: the universality of physical laws. But again, we are jumping ahead.

Never before has our universe been so open, actually expanding—and yet it is thoroughly closed in another sense. It is closed in the sense that whatever happens in this universe is steered by rigid laws of cause-and-effect. Since science wants to find out how this universe works, it strives to discover those underlying laws of cause-and-effect. No matter how you look at it, the universe appears to us as one big and intricate network of "cogs and wheels," with one cause set into motion by another cause, working with the

9. Supernovae are extremely luminous stellar explosions and cause a burst of radiation that often briefly outshines an entire galaxy, before fading from view over several weeks or months. During this short interval a supernova can radiate as much energy as the Sun is expected to emit over its entire life.

10. A black body is an idealized physical body that absorbs all incident electromagnetic radiation, regardless of frequency or angle of incidence.

precision of clockwork. No matter how you look at it, what we have here is one thing setting another thing into motion. That is how our world seems to work—as one long chain of events following one after another according to some specific laws of cause-and-effect.

No wonder the billiard-ball model has been for a long time standard in explaining the nature of the universe. It represents a kind of causal action that was thought to be evident, because the mechanism of this kind of action was supposedly clear and all-pervading. It was an example of "impulse," that is, of one body causing changes in another body by means of contact—by pushing it or striking it. "Impulse," the English philosopher John Locke once wrote, is "the only way which we can conceive Bodies operate in."[11]

Even our own bodies seem to be in the grip of some rigid clockwork mechanism. If we do not eat we die. If there is no oxygen we gasp for breath. If we drink too much alcohol our brains become affected. If we imbibe poison death sets in. Like it or not, everything in this universe—including you and me—is in the grip of some kind of rigid and fixed mechanism. The more philosophical term for this conception is *determinism*: Every cause has an inevitable and inescapable effect. As a consequence, everything would be determined by its past history, holding everything in the iron grip of determinism by working with clockwork precision.

Applied more in particular to human beings, mechanicism has become an important philosophical doctrine to declare that even all living objects, including human beings, are only and merely machine-like automata, which just follow all the physical laws of the universe, controlled by the machinery of their bodies. This idea is not new but is probably best known from the French physician and philosopher de La Mettrie who wrote a book entitled *Man a Machine* (1748). His ideas became part of a worldview called mechanicism.

Recent developments in science seem to support this view. The human body has become a "machinery" of biochemical pathways. The chemical model of biochemical pathways is a more recent version of the mechanical model of cogs and wheels. The metabolism of glucose, for instance, is a very deterministic, step-by-step process of breaking glucose down

11. John Locke, *An Essay Concerning Human Understanding*, II.8.11. He based this view on Isaac Newton's *Philosophiae Naturalis Principia Mathematica* (London, 1687).

into carbon dioxide and water, thereby releasing energy stepwise in small "packets" of ATP—and this is just one of the many intricate biochemical pathways at work in the body. All of this seems to be a matter of elaborate, cascading pathways of cause-and-effect. In this line of thought, Marvin Minsky, a pioneer in the field of artificial intelligence, described human beings as mere "machines made of meat."[12] Is that not what we had asked for…? Everything seems to be in the grip of determinism.

It should not be surprising that determinism is so attractive to us. It allows us, for instance, to rather accurately predict the positions of planets in our solar system at any point in time. The solar system consists of the Sun, the eight official planets, at least three "dwarf planets," more than 130 satellites of the planets ("moons"), a large number of small bodies (the comets and asteroids), and the interplanetary medium.[13] And here is the punch line: They all follow a predictable path according to the same basic laws—no surprises left.

Let us focus for a moment on the orbits of the eight planets of the solar system.[14] These orbits are ellipses, with the Sun at one focus—although all except Mercury are very close to circular. Thanks to these orbits, the position of specific objects in the solar system can be accurately predicted, because they follow very specific laws. The basis for the modern understanding of orbits was first formulated by Johannes Kepler whose results are summarized in his three laws of planetary motion. First, he found that the orbits of the planets in our solar system are elliptical, not circular, and that the Sun is not located at the center of the orbits, but rather at one focus. Second, he found that the orbital speed of each planet is not constant, but depends on the planet's distance from the Sun. Third, Kepler found a universal relationship between the orbital properties of all the planets orbiting the Sun.

---

12. Marvin Minsky, *The Emotion Machine: Commonsense Thinking, Artificial Intelligence, and the Future of the Human Mind.* (New York: Simon & Schuster, 2007).

13. This classification has become rather blurry, as there are several moons larger than Pluto and two larger than Mercury; there are many small moons that probably started out as asteroids and were only later captured by a planet; and comets sometimes fizzle out and become indistinguishable from asteroids.

14. Mercury, Venus, Earth, Mars, Jupiter, Saturn, Uranus, and Neptune (Pluto is now classified as a dwarf planet).

Then Isaac Newton demonstrated that Kepler's laws could be derived from his theory of gravitation. And next Albert Einstein was able to show that gravity was due to curvature of space-time, changing the gravitational field's behavior with distance; yet the differences from Newtonian mechanics are usually very small (except where there are very strong gravity fields and very high speeds). But no matter how you look at it, I would say this solar system was, is, and will be a very deterministic system—a system without any surprises.

The situation becomes a bit more ambiguous when it comes to forecasting the weather, but it is nevertheless of a deterministic nature again. Today's weather depends on what happened yesterday, and so will tomorrow's weather depend on today. If meteorologists give us a faulty forecast they can always blame something—the inaccuracy of the data, the complexity of the calculations, and so on—but never will they give up the idea of determinism. Science would not be possible if it did not assume that like causes have like effects and that the future depends on the past. In other words, the orderliness of this universe is actually a prerequisite for all science's endeavors.

However, if like causes always produce like effects, then the past appears to fully determine the future. It seems there is no way out of the grip causality has on us. So we end up with the doctrine of complete causal determinism, which the French astronomer Pierre Simon Laplace described emphatically as follows: "We may regard the present state of the universe as the effect of its past and as the cause of its future."[15] Science seems to confirm this over and over again: The future is completely determined by the past. Period!

Some might argue that science is discovering also a more *chaotic* element in this universe. Indeed, recently many scientists developed a new interest in chaos and chaotic systems, as if these could falsify the existence of "law-and-order." They point out that some natural systems can only be described by non-linear mathematical equations with such complex solutions that we cannot exactly predict what the system will do in the near future. Or to take another example, our measurements of all the initial

15. Pierre Simon Laplace (1749–1827), *A Philosophical Essay on Probabilities*, 6th edition, trans. into English from the original French by F. W. Truscott and F. L. Emory (New York: Dover Publications, 1951), 4.

conditions of a particular system (for example, in meteorology) may be too numerous and/or too inaccurate to predict what exactly the outcome would be. Even the movements of tiny butterfly wings may have a ripple effect on the weather. So the best we can come up with are probability statements.

However, this is not really chaos, but only appears to be chaos. As a matter of fact, scientists are still looking for the very order behind these seemingly chaotic phenomena. When the weather forecast is off the mark, we do not conclude that the weather is unpredictable, do we? Of course, we do not; we just do not know enough to be perfectly accurate. How would we ever be able to explain and predict at all if we were living in a world of mere chaos, disorder, and irregularity?

Let's face it, chaos is what it is—chaos! It is not some enigmatic form of "pre-order," but it is precisely what it says—chaos. C.S. Lewis once remarked, "Before you switched on the light in the cellar, there was (if you want to call it so) 'pre-light'; but the English for that is 'darkness.'"[16] Well, let me say it again: Chaos is not pre-order, but it is what it is—chaos! Chaos and chance can never create the order found in the living and nonliving world—just as blindness can never create sight. As the old saying goes, "what chance creates, chance destroys," because there is no purpose or direction to chance; if you do not believe this, just test it at a slot machine.

Instead it is the other way around: Chance is only intelligible in terms of the order which it lacks; a previous order must exist before any chance event can even occur. If there were no order, there could be no chance, because chance needs the order of preexisting causes coming together to produce unexpected results (which is called coincidence). Even the science of probabilities is based on very orderly probability distributions; it helps us to put a meaningful numerical value on things about which we do not have enough detailed knowledge; whereas a single random event may not be predictable, the aggregate behavior of random events is.

When events occur "by chance," it is not because events are uncaused or because we cannot trace their causes, but because so many causal chains independent of each other intersect. When people toss a coin, we just do not have enough detailed information (about force, angle, distance, etc.)

---

16. From Lewis' March 5, 1960 letter to Bernard Acworth.

to predict the outcome. But that is not the point; we do know that tossing a coin is subject to randomness because the outcome is independent of what the one who tosses the coin would like to happen, and it is also independent of previous and future tosses. Yet, in the aggregate, we are able to make predictions to a certain extent in terms of probabilities. But this is certainly not chaos. So seemingly "order-less" phenomena might very well turn out to be extremely orderly if only we had more detailed and more accurate knowledge. So let us not exchange the God of order for the deity of chaos.

Some might object that determinism in its all-pervading version is outdated and has forever been defeated by the physicist Niels Bohr and his school—in spite of objections made by giant opponents such as Albert Einstein, Max Planck, David Bohm, and Erwin Schrödinger. What is going on in this debate? According to Werner Heisenberg's 'principle of uncertainty' (or indeterminacy), it is impossible to determine simultaneously the values of position and momentum, or of energy and time, with any great degree of certainty; the more precisely one property is known, the less precisely the other can be known. Some objects just have multiple properties that appear to be contradictory. Sometimes it is possible to switch back and forth between different views of an object to observe these properties, but in principle, it is impossible to view both at the same time.

The question, though, is how to interpret this phenomenon. In Niels Bohr's interpretation, which is called the 'principle of complementarity,' these values are in essence un-determined until we measure one of them. He stresses "the impossibility of any sharp separation between the behaviour of atomic objects [i.e., objects governed by quantum mechanics] and the interaction with the measuring instruments which serve to define the conditions under which the phenomena appear."[17] Light, for example, behaves either as a wave or as a stream of particles depending on our experimental setup, but we can never see both at the same time. He assumes that these quantities have no precise values, so the behavior of atoms and electrons can presumably no longer be predicted until measured.[18] Would that be the end of determinism?

17. Niels Bohr, "Discussions with Einstein on Epistemological Problems in Atomic Physics." *Atomic Physics and Human Knowledge*. Science Editions, Inc. (1961): 39–40.

18. As Niels Bohr repeatedly said, "The quantum world does not exist."

I do not think we should give in that easily. Is it not ironic that Bohr and his Copenhagen school try to give us a *causal* explanation of the alleged fact that *causal* explanations are impossible in their view? They think that we *do* not know because we *cannot* know; they maintain it is impossible to regard objects governed by quantum mechanics as having intrinsic properties independent of determination with a measuring device. Furthermore, I would argue it is logically impossible to prove that something has *no* cause at all. Causality can never be conclusively defeated by experiments since causality is the very foundation of experiments.

No wonder that Einstein spent the last four decades of his life in a quest to restore order to physics. He realized that order does not come from science but actually enables science. He was right when he wrote, "But surely, a priori, one should expect the world to be chaotic, not to be grasped by thought in any way."[19] And yet, we seem to be able to grasp and comprehend the world to a certain extent. When the Nobel Laureates and astrophysicists Joseph H. Taylor and Russell Alan Hulse discovered the first binary pulsar, they showed, on the basis of tiny variations in the pulsar's radio emission, that the two stars are revolving ever faster and closer around each other at a rate agreeing precisely with what Albert Einstein had predicted in his relativistic theory of gravitation: Objects accelerated in a strong gravitational field will emit gravitational waves. The order we observe is amazingly very "consistent"!

Have we really lost contact with reality and causality since Bohr? I am not giving Einstein more credit here than he deserves, but I must confess that I tend to side with Einstein's philosophical qualifications, which easily surpass Bohr's. On the other hand, Bohr managed a 'well-oiled' public relations campaign for his interpretation and developed a powerful school of fans—including Heisenberg. But fans often just echo what they have learned from their master by hearsay.

Fortunately, who is right and who is wrong is not determined by majority votes. Come what may, the indeterminacy debate is still very alive in physics at the present time, and is apparently...well, indeterminate. Nevertheless, I keep betting on Einstein who so clearly realized that natural laws describe a reality independent of us; he was very aware of the fact that

---

19. Einstein's letter to Solovine, March 30, 1952.

order does not come from science but actually enables science. Order is not a scientific discovery but a scientific presupposition, making science possible.

## Living in an Evolving World

So far we have been discussing some recent developments in the science-world of physics and how these have affected our world-view. Well, it seems a similar touch of determinism has also invaded the science world of biology. It was already present before Gregor Mendel appeared on the scene, since most people thought then in terms of "blood" ties (terms such as "blood relative," "bloodline," "full-blooded," and "royal blood" are relics of this idea). So differences in temperament, talents, social status, wealth, and power were all said to reside "in the blood." With the rise of Mendelian genetics, however, genes were substituted for blood in the explanations.[20] That is when determinism became really rampant in the life sciences.

Genetic determinism tells us that the genotype completely determines the phenotype—which means that genes completely determine how an organism turns out. The simplest form of genetic determinism holds that the genes of parents inevitably determine all the characteristics of their children. A classical definition of a gene has it that a gene is a unit of heredity that determines a specific trait, feature, or characteristic of an organism. So there are genes that supposedly determine our blood types, our eye colors, and anything else in our lives. Since a gene can carry several variants, called alleles, all humans have the same genes, but not the same alleles—and these determine what we will be like, so they say.

How is this determinism supposed to work? Some genes work in one step; others need several steps to cause differences between people. There is, for instance, a gene that produces, in a *one*-step process, a certain enzyme, called *phenylalanine hydroxylase*. This gene can also harbor mutated alleles that produce a non-functional enzyme, which then may lead to phenylketonuria, causing mental retardation, seizures, and other serious medical problems when left untreated. This is a simple case of how a gene works. And there are a few more, such as a common form

20. As to whether Mendel was really the "founder" of genetics, see G.M.N. Verschuuren, *Investigating the Life Sciences.* (Oxford, UK: Pergamon Press, 1986), chapter 8.

of dwarfism, *achondroplasia*, which is caused by a mutated gene for a certain cell receptor (*FGFR3*) that leads to an abnormality of cartilage formation.[21]

But most other genes require additional steps to take effect. The gene for the ABO blood type can hold one allele for antigen A, one for B, or one for neither one of the two (O); these alleles produce an enzyme that then creates the corresponding A- and B-antigens. Here we have a *two-step* process. Another example would be a single gene that produces a certain enzyme (*tyrosinase*) which in turn creates the dark skin pigment melanin—again a two-step process. This gene can also harbor one or two abnormal alleles that produce a non-functional enzyme, which then may lead to albinism (but there are also other forms of albinism).

Then there are genes that require many more *steps*, or they even require the cooperation of many more *genes* in order to produce specific end-results. These genetic pathways may experience interference from various factors—and these interactions may even be of the nonlinear type. In addition, many genes interact with the environment, so we should rather speak of a genetic predisposition than of a genetic determination. A genetic predisposition means that an individual has a genetic susceptibility to developing a certain trait, but is not destined to develop that trait.

Actually, the development of many traits depends in large part on a person's environment and lifestyle. A simple example of the interaction between genetic predisposition and environment is the higher susceptibility of fair-skinned people to the development of skin cancer. In these lighter-skinned individuals, skin cancer can be associated with mutations in the melanocortin-1-receptor-gene (*MC1R*), but is also dependent on the length of exposure to sunlight. Another example of a genetic link to an environmental factor is seen among people with alcohol-dehydrogenase (ADH) deficiency caused by a mutation in the *ADH1C* gene, which encodes an ADH subunit. Only when they ingest alcohol, even in small amounts, ADH deficiency become manifest—otherwise it will not.

---

21. The *FGFR3* gene provides instructions for making a protein called *fibroblast growth factor receptor 3*. This protein is located on the cell membrane, and can interact with specific growth factors outside the cell. When these growth factors attach to the *FGFR3* protein, the protein is activated and regulates bone growth by limiting the formation of bone from cartilage. Mutations may cause the receptor to be overly active, leading to dwarfism.

There are many other complex traits that arise from numerous genetic and environmental factors working together. A good example is human height: It has a heritability of 89% in the United States, but in Nigeria, where people experience a more variable access to good nutrition and health care, height has a heritability of only 62%. Apparently, genetic simplicity has given way to complexity, and hence to a more complex kind of determinism—but it still is determinism in essence, although not of a simple kind, let alone of the mono-factorial type.

As a consequence, we remain at the mercy of genes and DNA, at least to a certain extent. For those who like to think in mono-factorial terms, the search is on for that one specific gene that determines some specific character in one's life. Someday, geneticists may discover, for instance, what that one fundamental genetic cause is behind all cases of Alzheimer's disease, or behind all cases of autism. In science, discoveries always start as inventions—usually called hypotheses.

However, not all inventions lead to discoveries. To use an analogy, the person who invented "Atlantis" did not discover Atlantis; it remains a legendary island until further notice. It is the same in science: Most inventions do not make it to discoveries. Yet some scientists think they have made a discovery when all they have in mind is an invention, a hypothesis. As a consequence, we have been bombarded with new genes: a gene for longevity, a gene for homosexuality, a gene for schizophrenia, even a gene for religion—the list could go on and on.[22]

Reacting to the casual way in which some neurobiologists speak of a gene for depression or a gene for violence, the British neurobiologist Steven Rose argues that depression and violence are merely simple labels for very complicated and variable patterns of behavior.[23] These hypothetical genes were once claimed, and then had to be retracted; they were often inventions that did not lead to discoveries. They were based on the mantra of mono-factorial determinism, but such a form of determinism only exists in the setting of simplified models (more on this later). Nevertheless,

22. The American geneticist-turned-filmmaker Dean Hamer, for instance, has recently postulated a "god gene" (*The God Gene: How Faith Is Hardwired into our Genes* (New York: Doubleday, 2004)), after he had conjectured a "gay gene" (*The Science of Desire: The Search for the Gay Gene and the Biology of Behavior* (New York: Simon and Schuster, 1994)).

23. Steven Rose, "The rise of neurogenetic determinism," *Nature*, 373 (1995): 380–2.

genetic determinism is still rampant in science, so hypothetical genes just keep coming and going. I would say that is where science borders science fiction.

Then science made the step from genes to DNA. Ever since the discovery of DNA, the boundaries of a gene have become rather fuzzy. The one-time idea of DNA holding only genes that code for proteins has been refined by the discovery that protein-coding regions of genes can be interrupted by DNA segments that play more of a regulatory role by producing proteins that either activate or repress the activity of "regular" genes. Some of these regulatory genes are actually very short and do not produce proteins at all but short strands of mRNA—called micro-RNAs—capable of blocking the mRNA of a "regular" gene from creating its protein. So the DNA picture has become more intricate than the gene picture, but it is still very deterministic. I will not go into further details here.[24]

Ironically, one of the DNA pioneers, Sidney Brenner, said not too long ago he could compute an entire organism if he were given its DNA sequence and a large enough computer.[25] With a like mind, the American molecular biologist Walter Gilbert had the audacity to claim, "that when we have the complete sequence of the human genome we will know what it is to be human."[26] I call that genetic determinism in full glory!

As a matter of fact, the ultimate goal of most scientists seems to be an explanation of life in terms of DNA; therefore, the "secret of life" is supposed to reside in DNA. In this view, all that counts is DNA; so with the human genome project finished, we are supposed to know all there is to know about human beings. DNA seems to have become the definitive controlling agency behind our lives.[27] That is the latest version of determinism—in denial of the fact that gene interaction and the impact of environmental and behavioral factors can be such that the whole (the phenotype)

---

24. See my upcoming book *It's All in the Genes! – Really?* (2014).
25. He made this claim at a symposium in commemoration of the 100th anniversary of the death of Charles Darwin.
26. Walter Gilbert, "A Vision of the Grail," Eds. D. J. Kelves and L. Hood, *The Code of Codes: Scientific and Social Issues in the Human Genome Project* (Cambridge, MA: Harvard University Press, 1991).
27. The science journal *Nature* listed "Personal Genomics Goes Mainstream" as a top news story of 2008.

is not only greater than the sum of its parts (the genotype), but the whole may be even different from the sum of its parts.

Recently, we have also learned how our genes and our DNA ever came along. Most scientists support the theory of evolution, especially in its neo-Darwinian version—and I admit being one of them. The defenders of this theory argue that all organisms descend from previous generations by gradual modifications, mainly based on a process of mutation (creating new alleles, and even new genes) and natural selection (favoring certain alleles over others).[28]

The neo-Darwinian theory of natural selection basically attempts to explain nature's beautiful design in terms of physical causes and natural laws. Darwin used to say that evolution follows laws in the same way as planets and comets follow laws in physics. The fact, for instance, that the caterpillars of a white cabbage butterfly are green rather than white (which is caused by alleles of certain genes), makes these slow organisms feeding on green cabbage less conspicuous to predators and thus more successful in survival and reproduction—which ultimately increases the frequency of these alleles in future generations. Biologists usually speak in terms of *functionality*: A functional design is a design successful in survival and reproduction.

Hence the reasoning goes as follows: Their green color is a successful or functional design because it causes camouflage, and therefore protection; as a consequence, natural selection causes this successful effect to spread. In other words, the green color of caterpillars has a selective advantage over other colors, and therefore increases its frequency through better chances of reproduction—it is a functional and successful design. In short, natural selection promotes good designs over bad designs, which makes them increase their representation in future generations. Natural selection promotes causes with successful effects. This seems to be "a law of nature" in Darwin's eyes.

Apparently, the life sciences deal not only with causality but also with functionality. But make no mistake, functionality is not at odds with

---

28. This is often called "common descent with modification." In his book *On the Origin of Species by Means of Natural Selection*, Darwin did not use the word "evolution," but always referred to it as "descent with modification."

causality. Functional relationships are always causal, but causal ones are not always functional. The relationship between heart beats and heart sounds is certainly causal—beats produce sounds—but it is not a functional relationship—the sounds do not benefit survival. The relationship between heart beats and blood circulation, on the other hand, is not only causal—the heart beats make the blood circulate—but also functional, for the beats have a successful effect—the beats are there "in order to" have the blood circulate.

This leads us to the following question: What is the causality behind this functionality—given the fact that we are supposed to live in a world of "law-and-order"? Since Darwin, the answer would be that the caterpillars we mentioned before are green, as a population, because of two different types of causes. They are individually green due to some particular genetic mechanism—which is sometimes called a "proximate" cause. In addition, they are green as a population because of the fact that their green ancestors successfully deceived potential predators in the past—which is often called an "ultimate" cause. The green color is a cause that makes for a functional biological design and hence will be promoted by natural selection.

It is the latter type of causes that is important in evolutionary biology. Selective reproduction is the causal explanation of how the green color of certain caterpillars could become so widespread in the entire population or species. The green color "made" it in evolution among these animals, because it is a successful and functional design in the environment they live in. Natural selection selects what is functional by favoring causes that have "successful" effects. In evolutionary biology, survival has no significance unless it adds to reproduction—success breeds success. Dayflies, for instance, may not survive long but they do reproduce well, whereas mules do not reproduce well, no matter how excellent their survival may be. So functionality is somehow causality tested in the furnace of evolution, generation after generation.

This explains why causality does, but functionality does *not*, occur in the physical sciences, for the latter do not deal with selection based on reproduction. When the temperature rises, gases do not expand "in order to" keep the pressure constant. There is no functionality here. But hearts do beat "in order to" keep the blood flow going. A beating heart is a good functional design, and natural selection favors causes that have "successful" effects. So the causal relationship between heart beats and blood circulation is functional, whereas the causal relationship between heart beats and heart sounds is not, since the latter connection does not contribute to the success of an organism in surviving.

All of this seems to fit nicely in a world of "law-and-order." Based on this conception, the neo-Darwinian theory assumes a process of gradual change—step by step, generation by generation, species by species—but on a genetic basis—gene by gene, mutation by mutation. How could such a process give rise to a species that is per definition reproductively isolated from other species? To state it more powerfully, how can a species ever change into a new species, given the fact that all species are reproductively isolated from each other? If there is such a process—and this seems to be more and more likely—it would be called speciation.

Usually speciation starts with some form of geographical isolation. Due to accumulating mutations, geographical barriers lead in time to biological barriers. Take, for instance, the two fruit fly species *Drosophila pseudoobscura* and *D. persimilis*.[29] They are very closely related, yet isolated from each other by habitat (*persimilis* generally lives in colder regions at higher altitudes), by the timing of their mating season (*persimilis* is more active in the morning and *pseudoobscura* at night), by behavior during mating (the females of both species prefer the males of their respective species), and by sterility of hybrid males. Even if the original physical barrier no longer exists, the two new species may stay reproductively isolated due to their acquired biological barriers.

Once geographical isolation has created reproductive isolation, speciation has become a fact. The situation is comparable to the spread of English over the earth. People of England, Ireland, India, Canada, USA, and Australia all speak English, but while they were more or less isolated they developed their own "dialects." Do they still speak the same language? In a way, they do—in another way they do not. Had the isolation lasted much longer, they could have developed their own language and would probably no longer be able to understand the "English" of people isolated from them.

But there is more. Not all geographical isolation is based on physical

29. Carlos A. Machado, Richard M. Kliman, Jeffrey A. Markert, and Jody Hey, "Inferring the History of Speciation," *Molecular Biology and Evolution*, Volume 19 (2002), no. 4: 472-488.

barriers such as mountains, deserts, or water masses. There are also "barriers" of a different nature—such as the mere physical distance between some members of a population. In such cases, the population is still continuous, but nonetheless, its members may not mate randomly, because they are more likely to mate with their geographic neighbors than with individuals at a greater distance.

Let us consider the case where the species forms a ring of neighboring populations who can interbreed with adjacent populations. There may be at least two "end" populations in the series that are too distantly related to interbreed. An example of this situation can be found in California, where the *Ensatina* salamander forms a horseshoe shape of populations in the mountains surrounding the Californian Central Valley.[30] Although interbreeding can happen between each of the nineteen populations around the horseshoe, the salamanders on the western end of the horse-shoe cannot—or can no longer, I should say—interbreed with the salamanders on the eastern end.

The question, then, arises whether to classify the whole ring as a single species (in spite of the fact that not all individuals can interbreed) or to consider each population as a separate species (in spite of the fact that there is still interbreeding with nearby populations). This much is clear: If enough of the connecting populations within the ring perish to sever the breeding connection, the remaining members would have become two distinct species. It should not surprise us then that, in an evolutionary context, a clear-cut species concept inevitably becomes a bit fluid. The situation is again comparable to what happened to English in the USA. People from the North may have some difficulty understanding people from the South, since they have developed their own dialects being so far away from each other. But they have not made it (yet) to different languages.

Then we have at least one more problem left: A mutation can create new alleles for existing genes, but how could a mutation ever lead to *new* genes? Let me explain this with the way a modern biologist would assume a step-by-step scenario of a selection process for the intricate cascading route of human blood-clotting, which involves quite a few proteins (serum proteases).

---

30. T. Dobzhansky, *A Century of Darwin*, ed. Barnett S A (Cambridge, MA: Harvard Univ. Press, 1958), 19–55.

Most of the proteins in the blood-clotting cascade turn out to be related to one another at the level of amino acid sequence, which most likely reflects ancient gene duplications.[31] Since new copies were not essential for the original function, they could gradually evolve to take on a new function, driven by the force of mutation and natural selection. And indeed, we do have evidence for such a process: Some of these proteins have a long history in the animal world, whereas others did appear much later and made the cascade of clotting factors more intricate when it evolved from a low-pressure to a high-pressure cardio-vascular system based on a more extensive series of proteins.

There is a similar story for the human beta-globin cluster as part of hemoglobin formation in red blood cells. It is composed of five genes located on a short region of chromosome 11. The sequences of these genes are quite similar, which suggests they occurred by duplication of an ancestral beta-globin gene. Together all five functional genes are responsible for the creation of the beta parts (roughly half) of the oxygen transport protein hemoglobin but they do so at different moments during development: The epsilon gene is expressed during early embryo development, the two gamma genes during fetal development, the delta gene early after birth, and the beta gene throughout the remainder of the life cycle. The arrangement of the genes next to each other on the chromosome directly reflects the temporal differentiation of their expression during development. If the genes are rearranged, the gene products are expressed at improper stages of development.

Let us come to a conclusion. There is increasing evidence for the process of speciation; it is enforced by geographical or other forms of isolation and leads to accumulated variation steered by natural selection. Are there dissenters regarding the neo-Darwinian theory? Of course there are; every field in science has its dissenters. In time, we may find a better theory, but currently this seems to be the best we have. However, if this theory is true,

31. Genes that have a similar sequence of DNA building blocks (nucleotides) and produce very similar proteins are classified into "gene families." Gene families are a strong indication that the DNA-sequences for these proteins are duplicates of each other, but with some minor changes (mutations). Examples are blood-group-antigenes, immunoglobulines, and oxygen carrying globins (hemoglobin in blood and myoglobin in muscles).

we end up living in a world that evolves through random events such as mutations, combined with a good dosage of deterministic processes. We need to find out if this is the end of the story, or if there is more to it.

## Running with Clockwork Precision

Whether it is in physics or in biology or in any other field of science, everything appears to be in a process of change according to preset laws of nature. It was in 1872 that the British philosopher John Stuart Mill promoted a rule that says "it is a law that every event depends on some law."[32] Whatever happens by the force of a "law of nature" is something that *must* happen in this world of ours, and we speak of things covered by such laws as "natural." All processes in nature seem to follow scientific laws—physical, biological, or what have you. It is these laws that control everything in the universe, including the path of electrons, the motions of planets, as well as the evolution of species.

All of this leads us to believe that we seem to be living in a world ticking with predetermined clockwork precision—and science wants to explain how this mechanism works. If there were merely change in the universe—as the Greek philosopher Heraclitus once claimed—explanation would be senseless. On the other hand, if there were only stability—as the Greek philosopher Parmenides used to say—explanation would be needless. Explanation means that we perceive an underlying order in spite of apparent chaos and constant change.

Explanations are usually done by invoking a general law with the following very simplified structure: If X, then Y. Consider a law like this: If the rate of metabolism goes up (X), then the rate of heart beats goes up (Y). With such a law, we are able to explain and to predict: We can *explain* Y with the occurrence of X, or we can *predict* Y when X happens. That sounds very rigid, does it not? However, laws only work within a series of boundary conditions. The law stating that water boils at 100° Celsius is only valid if the water is pure, the pressure is normal, etc.

This is one of the reasons why it is usually much easier to explain than to predict, since we do not have to know all boundary conditions in order to

---

32. John Stuart Mill, *A System of Logic* (Honolulu: University Press of the Pacific, 2002) chapter V, § 1.

give an explanation. I would even add that prediction is still more difficult when the *future* is concerned—but that is meant as a joke. When making predictions, we may not be sure which conditions are relevant and to which extent. On the other hand, sometimes it is easier to predict than to explain, especially when it comes to so-called correlation laws; a strong decline in barometer level is correlated with an upcoming storm, but that decline is not the cause of the storm and does not explain the storm; the storm cannot be explained by mentioning an additional, correlated effect of this cause. We all know that the rooster's crow does not *cause* the sun to rise.

Another problem is that sometimes a cause only refers to the *sufficient* conditions of an effect—conditions in whose presence the effect is bound to occur.[33] Here are a few examples: Lack of oxygen always causes brain damage; the toxic protein ricin always blocks protein synthesis; large amounts of alcohol always cause intoxication; high temperatures always cause proteins to denaturate; the presence of a certain allele of the *HEXA* gene always causes Tay-Sachs disease.[34] Notice that in all these examples the word "always" may be too optimistic; under certain conditions the effect may *not* occur.

At other times, the cause specifies only the *necessary* conditions of an effect—conditions in whose absence the effect cannot occur. Here are some examples: Alcoholism cannot occur without the consumption of alcohol; chickenpox cannot occur without the presence of the *VZV* virus; the enzyme alcohol dehydrogenase cannot work without the presence of zinc. These conditions are necessary but not necessarily sufficient.

Usually, however, scientists tend to select one, or at most a few, conditions as *the* cause; hence all other conditions are considered "background" conditions. When geneticists take a certain genetic factor as the cause of a certain disease, they are not denying the presence of other factors, such

---

33. The dichotomy between sufficient and necessary causes is rather fuzzy; there are many borderline cases and combinations. This distinction is also used in logic for a relationship between statements or propositions. When applied to the relationship between events, however, there is a temporal order involved: The second event (the effect) is taken as a consequence of the first event (the cause).

34. Some may question whether the word "always" in the last two cases is correct. One could make the case that we do not know whether there are people walking around with that same allele who never developed the disease.

as food and oxygen, which are also necessary conditions for this outcome. Obviously, for the sake of simplicity, they only focus on those causes "that make the difference," while leaving out all other necessary conditions. An additional problem is that we often speak of sufficient conditions without realizing that certain necessary conditions are so prevalent that they easily elude attention.

So when talking about scientific laws, or laws of nature, we must always be aware that such laws operate in a model-like setting that leaves certain conditions unspecified. For that reason, scientific explanations and their laws usually only refer to sufficient conditions, while leaving out the necessary conditions of the boundary conditions that come with the *model* that was used. No wonder then that it is very hard to find good examples of sufficient causes, because often there are situations where the unspecified boundary conditions do not hold. So what could we learn from this? Scientists always work with model-like settings—to put it briefly, they use models!

Well, the fact that scientists can select one or more causes to the exclusion of others has some important consequences. Any explanation can be based on a variety of models—that means, on various kinds of background conditions—which makes it possible to explain a certain phenomenon with reference to a wide range of causes. In other words, the frame of reference may vary quite a bit, so one can look at certain phenomena from various angles.

Let me explain this with a simple example—mating behavior in the animal world. This behavior is a phenomenon-in-general which has several causes-in-specific, depending on the model that is used. An ecologist may point to certain seasonal factors causing particular physiological changes in the female animal, whereas the causes mentioned by physiologists would be of a different nature: A neuroscientist may refer to nerve impulses and transmitter substances between neurons, whereas an endocrinologist would speak of certain hormones activating their target organs. These explanations differ significantly from those given by a geneticist, who would locate their causes in certain genes carrying the hereditary code for this behavior. And last but not least, an evolutionist may look for causes located in a long evolutionary process. To put it simply, they all work with different models which are based on different assumptions and refer to different causes and different boundary conditions. Models always *select* and therefore *neglect* what is not selected.

As a consequence, there is no simple answer to the question of what is the cause of something like AIDS. The answer depends on the model we use. According to the model of virology, it is the human immunodeficiency virus (HIV) that causes the disease. However, this virus does not always cause the disease, because the model of human physiology allows for other factors to interfere, which may lead to different degrees of resistance and a variety of symptoms. Next the model of epidemiology tells us that the chance of developing AIDS is the highest in South Africa. In a behavioral model, on the other hand, factors such as unprotected sexual intercourse, contaminated blood transfusions, and the use of hypodermic needles can cause transmission of the virus. Finally, the model of evolutionary biology indicates that the HIV virus originated in southern Cameroon through the evolution of another virus (SIV) that infects wild chimpanzees. Apparently, each model focuses on a different aspect of the disease and speaks of different causes.

The question as to which causes are relevant just depends on the model we are working with and on the problem our research efforts are trying to solve. In other words, there is no definitive explanation for and no decisive cause behind anything that goes on in this world—in spite of the fact that most scientists tend to think that *their* explanation is *the* explanation, instead of taking it as an answer to a very specific question posed within the setting of a very specific model.

To put it differently, all specialists have their own little "story" to tell, a story that lacks completeness without other "stories." Obviously, our world is so complicated that one story could never do the job alone. The cause of an epidemic in a harbor city, for instance, may be regarded by bacteriologists as the microbe they find in the blood of the victims, by entomologists as the microbe-carrying fleas that spread the disease, by epidemiologists as the rats that escaped from the ship and brought the infection into the port. Who has the final answer? None of them does, but together they may have a good story.

At some point in my life I came to realize that all explanations based on scientific laws are actually strange entities—basically of a "hybrid" nature.

They explain material things and events by using *non*-material *laws* of nature. What are immaterial laws doing in a world of material entities? That is rather odd, for laws are unlike anything else in this material world. Right in the middle of our comfortable, spacious, temporal, transient, and piecemeal world of causes and effects, something pops up that we call "laws of nature"—physical and biological laws, mathematical laws, and even moral laws.

Unlike all material things surrounding us, laws do not have any of the features that apply to the material world—that's right, *none*. A law is not located somewhere in space, not even in our minds, for the mind just has a mental picture of the law—but a law holds everywhere. It is also beyond time, a timeless entity that cannot emerge nor perish in the history of the universe. Neither are laws subject to change, for they will always remain true, even before we discovered them. And here is the most important difference: Not only are laws general and universal, as we find the same law applied all over the universe, over and over again, but also are they necessary, which means that things in this universe cannot be different from what is expressed in laws.

Apparently, the inhabitants of this immaterial world are completely unlike the inhabitants of the material world. Laws are "unseen," very different from all that is "seen" around us. They belong to a different world—a "virtual" world of thoughts and truths so to speak, and yet a very "real" world. Unlike the things that can be "seen," laws are not three-dimensional, temporal, transient, individual, and contingent. They almost look like ghosts from the outer-world! In a physical sense they are nowhere, and yet they are everywhere. Would it not be much simpler if we could make them disappear, or transform them, so they become fundamentally identical to all other regular things in this physical universe?

It should not surprise us then that some thinkers, such as the Scottish philosopher David Hume, have seriously tried to get rid of laws by claiming that they are just mental creations: "We have no other notion of cause and effect, but that of certain objects, which have always conjoin'd together, and which in all past instances have been found inseparable."[35] In his view, the

35. David Hume, *A Treatise on Human Nature* (1739-40), ed. L. A. Selby-Bigge (1888), book 1, part 3, section 6, 93.

basis of our concept of causation is merely the *mental* act of association.[36]

If Hume were right, laws would exist *only* in the minds of mathematicians, physicists, biologists, and ethicists. However, this solution fails to explain the fact that laws actually do hold in the real world.[37] How is it possible for a bridge that has been designed according to the right laws to stand firm, whereas another bridge collapses because its engineers erred in their calculations or used the wrong laws? Would competent engineers really have better mental habits than their inept colleagues? So my question is: How could these laws ever hold if they were only creations of our minds?

Therefore, I would argue that laws cannot easily be bypassed in the way Hume suggested. They are not something material, a brain wave, but immaterial—call them mental, spiritual, or whatever you prefer. I am reluctant to call them mental as that adjective might suggest they exist only in our minds à la Hume. On the contrary, they are very "real" and exist outside the mind. We may know them but we cannot create them; they are there even before we know them. That is the reason why laws of nature have to be discovered, not invented. Without such "given" laws, our minds would not be able to tell true from false and right from wrong.

Interestingly enough, science is discovering more and more that the universe has an inner logic accessible to human reasoning—an intrinsic rationality that governs all that science is trying to decipher. Or to quote a famous scientist, the late astrophysicist Sir James Jeans, "the universe begins to look more like a great thought than a great machine."[38]

So the "law-and-order" framework of the universe still stands up straight—at least to a certain extent. We seem to be living in a closed world, ruled by "law-and-order," similar to a windowless prison cell. Everything seems to be working autonomously, ticking with clock-work precision. By "closed

36. "From causes which appear similar we *expect* similar effects. This is the sum of our experimental conclusions." David Hume, *An Enquiry Concerning Human Understanding* (1748), ed. L. A. Selby-Bigge (1894), section 4, part 2, 36.

37. "That the sun will not rise tomorrow is no less intelligible a proposition and implies no more contradiction than the affirmation that it will rise." David Hume, *An Enquiry Concerning Human Understanding* (1748), ed. L. A. Selby-Bigge (1894), section 4, part 1, 26.

38. James Jeans, *The Mysterious Universe* (Cambridge University Press, 1930), Chapter 5.

world," I do not mean a static world, for our universe appears to be a universe in expansion and in evolution according to the latest theories. By "closed" I mean: There is nothing "outside," "beyond," and "behind" the universe, beyond and behind all we can see through our telescopes and microscopes. Everything works according to its own rules. That is how I came to view the world during my training as a scientist.

But now I realize I should put this more carefully for now: All we can say so far is that there *seems* to be nothing and no one out there— nothing and no one beyond the workings of the universe. But appearances may be very deceiving!

# 2. An Opening in the Fence

In the previous chapter I kept repeating: "There seems to be…?" We *seem* to be living in a closed world. We *seem* to be surrounded by a purely material world. We *seem* to be genetically determined. We *seem* to be living in a world without purpose. And so on and so forth. That is the impression science gives us—or at least that is the picture some scientists paint us. What is missing in this picture? A lot, as I have gradually found out. I would point at a whole world of immaterial things that eludes us—things such as laws, as well as thoughts, values, beliefs, emotions, hopes, dreams, and ideals.

Why are we missing out on so much? Could it be because we are looking through a scientific lens, which might distort what is out there? We discovered already a small opening in the seemingly closed scientific fence, when we discussed how science tries to deal with the material world of planets, genes, and molecules in terms of im-material laws of nature. That might be our first opening, our first way out.

## Multiple Windows

Most scientists love their work. The Nobel Laureate and astrophysicist Anthony Hewish once remarked that "scientists have a duty to share the excitement and pleasure of their work with the general public." That having been said, there are also many scientists who suffer from megalomania and tunnel vision. They claim to have all the answers to all possible questions—including the answer that science is all there is and all that counts.

In response, I would give them this warning: You may neglect—but cannot deny—what is beyond your horizon. If you do, you have a very myopic view on the world, declaring everything else in life outside your narrow scope as a mere illusion. Scientific models always select and therefore neglect what is not selected. Yet, some scientists claim that, as scientists, they know everything about the universe, based on some esoteric knowledge that everyone else supposedly lacks. The physical chemist Peter

Atkins even had the arrogance to exclaim that "scientists … are privileged … to see further into the truth than any of their contemporaries."[1]

People like Atkins somehow fail to remember they are just specialists like any other specialists; they are specialists in doing scientific research regarding the material aspects of this world—physical, biological, or whatever—leaving everything else for other "specialists." If they claim expertise in everything else as well, they are like plumbers trying to fix your electricity—or electricians fixing your plumbing. Expertise in one field may not help one handle problems in another field. So science is not what it is cracked up to be in the minds of some of its arrogant fans.

Many scientists seem to have forgotten that society once gave them their own "territory" to work in. In 1662, when scientific societies began to blossom, the fellows of the Royal Society of London received from King Charles II the privilege of enjoying intelligence and knowledge, but with an important, crucial stipulation in their charter of 1662: "Provided nevertheless, that this our indulgence […] be not extended to further use than the particular benefit and interest of the aforesaid Royal Society in matters of things philosophical, mathematical, or mechanical."[2] Notice there is an explicit limitation in the phrase—"in matters of..." Everything else was left for others to handle, as long as their own members, the future scientists, were permitted to "know" what can be known by counting and measuring.

Robert Hooke's draft of the Royal Society statutes reads literally: "The Business and Design of the Royal Society is: To improve the knowledge of natural things, and all useful Arts, Manufactures, Mechanik practices, Engyries and Inventions by Experiments—(not meddling with Divinity, Metaphysics, Moralls, Politicks, Grammar, Rhetorik, or Logic)."[3] Again there is a clear limitation here: "not meddling with…" That is how the

1. Peter Atkins, "The Limitless Power of Science," *Nature's Imagination: The Frontiers of Scientific Vision*, eds., John Cornwell and Freeman Dyson (Oxford: Oxford University Press, 1995), 121.

2. http://royalsociety.org/uploadedFiles/Royal_Society_Content/about-us/history/Charter1_English.pdf (page 11). "Philosophy" is not understood here in the modern sense but in the Newtonian sense of "natural philosophy"—the study of nature and the physical universe.

3. George S. Emmerson, *Engineering Education: A Social History* (Crane, Russak & Company, 1973), 19.

"division of the estate" was executed. By accepting this separation, science bought its own territory, but certainly at the expense of inclusiveness. The rest of the "estate" was reserved for others to manage. That is where faith, religion, ethics, and philosophy would come in; they should forever remain part of a different territory, beyond the scope of science, and therefore outside its reach.

When the first scientists, later followed by the Royal Society of London, decided to clearly fence off their area of investigation—by limiting themselves to questions that could be answered by material, physical, and mathematical means—they created their own framework of reality, which does not mean that the territory they left untouched is not equally "real" and important. They were allowed to explore the "Book of Nature," but things such as the "Book of Scripture" were off-limits for them, left for others to manage.[4] Ever since, all scientists have been profiting from this demarcation but, unfortunately, often have forgotten what it entails—limitation. A good exception is the physicist William D. Phillips, who ended his acceptance speech for the Nobel Prize in 1997 with these words: "I thank God for providing such a wonderful and intriguing world for us to explore."[5]

I would say in reply to those who claim scientific superiority that science certainly does not know all or cure all. I know of a Jesuit biologist who used to tease his parishioners and challenge his students with a quip: "You don't need to tell me anything about life—I am a biologist." Coming from his mouth, it was a joke! But in the mouths of some scientists it is not! Students in our schools—from elementary school to college—deserve to be taught genuine science, so they and their parents should not settle for some kind of ideology. Hence, in teaching science, we must also make clear what its limitations are—such is part of teaching genuine science as well. Teach it, but do not preach it.

The late Harvard biologist Stephen Jay Gould put it this way: "The magisterium of science covers the empirical realm: what the Universe is made of (fact) and why it works in this way (theory). The magisterium of religion

4. This distinction can already be found in the writings of Augustine: "It is the divine page that you must listen to; it is the book of the universe that you must observe." (*Enarrationes in Psalmos*, 45, 7).

5. William D. Phillips, "Laser Cooling and Trapping of Neutral Atoms," Nobel Lecture, December 8, 1997.

extends over questions of ultimate meaning and moral value. These two magisteria do not overlap."[6] All attempts of merging the two into one concoction are like trying to create a mixture of "oil and water"; they just do not mix well together. So keep them separated, for good fences usually make good neighbors. Yet, both sides must acknowledge there *are* fences; the ideology of scientism, on the other hand, does not acknowledge any neighbors and tries to silence them. It claims to have all the answers to all possible questions, including the answer that science is all there is and all that counts. That is why scientism is basically an ideology (more on this later).

We discussed already in the previous chapter how scientific explanations always feature within a specific range of reference. A geneticist, for example, works in a framework very different from a physiologist—and yet they may even study the same organisms. Once we leave the area of biology for the area of physics, the range of reference changes even more dramatically. A more technical terminology has it that the various scientific fields work with different *models*—or maps, if you want—of this world. A genetic "map" is different from a physiological "map," not to mention an astronomical map; they were developed in different frameworks based on disparate models. All models select, and thus neglect what they did not select. That is why it does not make sense trying to find cities or organisms on astronomical maps, but this does not mean they do not exist.

Earlier we discussed that a scientific explanation and its laws usually only refer to sufficient conditions, while leaving out the necessary boundary conditions that come with the model that was used. Apparently, a model is a simplified replica of the dissected original, made for research purposes. Every model deals only with one specific aspect of the world, so it does not deny other aspects but just ignores them. Hence, there are many useful and truthful "maps" of this world, but each one of them is, by nature, only a surrogate for "the real thing"—with everything else left out of the picture. Each "map" offers us only a limited, "one-sided" view and interpretation of reality. That is why we can have geographic maps, economic maps, genetic maps, neurological maps, and so on—they do not compete with each other but at best complement each other.

Let me stress it one more time: Do not think that a simplified model

---

6. Stephen J. Gould, *Rocks of Ages* (New York: Ballantine Books, 2002), 207–208.

with its simplified map is "the real thing." Do not declare the simplified model to be the new reality. If you do, you would be ripped off with a stripped version of reality. Never mistake a scientific map or model of the world for the world itself, since a map is only a surrogate for "the real thing." The only model that could ever qualify as a perfect replica of the original is the original itself.

As a consequence, there is no "science of the entire world," for each scientific field has its own outlook on the world, thus creating its own phenomena and its own facts. A chemist has an eye for chemical phenomena, whereas a biologist perceives biological facts. Consequently, each scientific field creates its own map of the world, building its own simplified and reduced version of the real world. Once you realize this, you would never replace the real world with one of its scientific maps or models. The fact that highways are missing on railroad maps does not entitle us to deny their existence—at least we should know better.

An added problem is that our scientific maps are getting more and more detailed, because science is becoming more and more specialized. However, "more detailed" does not mean "more comprehensive." We may make more and more maps that may convey more and more detailed knowledge, but we seem to know more and more about less and less. Besides, those more detailed maps are less and less connected. Our uni-versities have become multi-versities, a vast cluster of islands, with each one having its own specialty.

Is it not amazing how scientists wish to zoom in into finer and finer details, up to a square nanometer or micron, but then tend to lose sight of the wider context? More specialization usually leads to a narrower vision, rather than a more general overview. Models can certainly enhance our view—with a better focus, a higher precision, and more details—but at the same time they can also narrow our view by obscuring what is outside their scope. Ironically, science gives and science takes away at the same time.

It is becoming quite obvious now that *facts* have something to do with the kind of model we use. The physicist and philosopher of science Karl Popper used to tease his audience with the command "Observe!"—which necessarily evokes the question "Observe what?"[7] There is no observation

7. Karl R. Popper, *Conjectures and Refutations: The Growth of Scientific Knowledge* (London: Routledge Classics, 1963/2002), 61.

without some form of expectation—which entails interpretation. Take a surveillance camera: it "observes" everything because it does not know *what* to observe.

The problem we have here is best expressed by the good old philosopher Plato, although in a slightly different context, when he said: "How would you search for what is unknown to you?" Plato noticed a seeming paradox here: We are in search of something "unknown"—otherwise we would not need to search anymore. And yet it must be "known" at the same time—otherwise we would not know what to search for, or would not even know if we had found what we were searching for.[8] This is the reason why we do need hypotheses in science, as those can open our eyes for *facts* we would not have been able to see without them.

What then are those famous "facts"? Francis Crick, one of the two scientists who discovered DNA, could not have said it better: "A theory that fits all the facts is bound to be wrong, as some of the facts will be wrong."[9] How could facts ever be wrong? Are they not a matter of fact? Crick is right: What is "out there" is not an assortment of rock-solid *facts*, but rather a collection of things and events; facts are merely our *interpretations* of those things and events, our way of making them intelligible for us.

Events may be the "physical" parts of our world, perhaps even rock-solid, but facts are "mental" creations—the interpretations of events in our minds.[10] In observation, one is both a passive "spectator" and an active "creator" at the same time.[11] We look at *events* as spectators, and then change

---

8. *Meno* (80d). It is in an ethical context that Plato poses this question to Socrates: "How would you search for what is unknown to you? For, which idea would you have of what you are looking for, while it is unknown to you? Even if you would come across it, how could you know that it is the same as what you did not know?" This is used as a key passage in: Th. Kisiel, "Zu einer Hermeneutik naturwissenschaftlicher Entdeckung," *Zeitschrift für allgemeine Wissenschaftliche Theorie*, 2 (1971), 203.

9. Francis Crick, *What Mad Pursuit: A Personal View of Scientific Discovery* (New York: Basics Books, 1990).

10. If you want a comprehensive definition of a fact, try this one: A fact arises at the intersection of events, thoughts, and statements. It is the description of an event, the object of a thought, and the content of a statement. It is through *interpretation* that thoughts and statements transform events into facts.

11. A simple example may help explain this. Describing movements in the sky by saying "Those are moving *spots*" conveys rather empty information with nearly zero

them into *facts* as creators. We do not "have" observations—like we have sensorial experiences—but we "make" observations. This is the reason why a camera, for instance, cannot capture facts—all it can capture is things and events. So it should not surprise us that cameras cannot replace scientists! The problem with pictures is that they do not show us facts until we give some interpretation to what we see on the picture. The same with books: They provide lots of information for "bookworms," but to real worms they have only paper to offer. Facts are man-made. Every fact is interpreted within a specific framework, within a specific model if you will, or—to put it differently—every fact is seen from a particular perspective.

Let me illustrate this with the famous historical case of the astronomer Galileo Galilei. The scientific case Galileo was promoting was not as clear as many think, because the astronomical "facts" were not very clear. First of all, when Galileo supported Copernicus' heliocentric model[12] in his 1632 book called *Dialogue Concerning the Two Chief World Systems*, he had to deal with some difficult "facts." If we go by Aristotelian theories of impulse and relative motion, the theory advanced by Galileo appears to be falsified by the "fact" that objects appear to fall vertically on earth rather than diagonally—the famous so-called tower argument.[13] Other "facts" seemed to confirm as well that the earth did not move, for if it did, the clouds would be left behind—a "fact" that Galileo himself had already remarked in a lecture of 1601. As Paul Feyerabend, the late University of California at Berkeley philosopher of science, pointed out, one could even state that Galileo's opponents kept closer to the "facts" than Galileo himself.[14]

The observation that objects fall vertically on earth required a new *interpretation* to make it compatible with Copernican theory. Galileo

---

interpretation; by saying "Those are flying *birds*," one claims more and therefore has to prove more; by saying "Those are migrating *geese*," one includes even more interpretation. Only research can tell us whether each interpretation is correct.

12. Copernicus' heliocentric model was based on circular orbits, whereas Kepler's model worked with elliptical orbits. Galileo just was not particularly interested in the fine details of celestial mechanics, and 8' of arc errors were not his main concern.

13. However, the tools of observation and the measurements at the time were not accurate enough to decide this issue.

14. Paul Feyerabend, *Against Method: Outline of an Anarchistic Theory of Knowledge* (New York: Verso Books, 1975), chapter 13.

was able to make such a change about the nature of impulse and relative motion, but before such theories were articulated, he had to use ad-hoc methods and proceed counter-inductively—given the knowledge available at the time. Galileo even spoke of "a rape of the senses," when he denied the rule of his eyes and all previous conclusions about moving objects. In other words, Galileo had to "create" his own "facts."

This takes us to a related question: How reliable was Galileo's "tube" with which he created and confirmed his new "facts." His first telescope was based on an existing optical device in Holland, called spyglass, whereas modern refracting telescopes are based on a design proposed by Johannes Kepler in 1611 but first constructed by the Jesuit Christopher Scheiner, somewhere between 1613 and 1617.[15] However, just like students who use a microscope for the first time and see hardly anything, so astronomers must learn to use telescopes too. When Galileo demonstrated his simple telescope to a group of professors in Bologna in 1610, all admitted the instrument seemed to deceive; some fixed stars were actually seen double. Galileo had to concede in a letter to Johannes Kepler that many people were unable to see what they were "supposed" to see through his telescope.

Ironically, even Galileo himself would refer to comets as "optical illusions," rather than "facts," when he thought it would suit him well during his dispute over comets with the Jesuit astronomer Horatio Grassi—a dispute Galileo would eventually lose. We must also realize that there were no real experts in the field of optics at the time, except for Johannes Kepler and Horatio Grassi (and later on, scientists such as René Descartes, Isaac Newton, and Christiaan Huygens). So, understandably, many scholars thought that all the things the new telescope showed them could only be "arti-facts," not "real-facts." The German playwright Bertolt Brecht has one of the characters in his play, the mathematician, remark to Galileo, "One might be tempted to reply that your telescope, showing something which cannot exist, may not be a very reliable telescope, eh?"[16] The world of "facts" is certainly not a rock-solid world.

The fact that facts are always interpreted from a specific perspective

---

15. Albert van Helden, "The Telescope in the Seventeenth Century," *Isis*, Vol. 65, No. 1, 1974.

16. Bertolt Brecht, *Galileo*, trans. Charles Laughton, ed. Eric Bentley. *Works of Bertolt Brecht* (New York: Grove Press, 1966), 43–129.

does not only apply to the variety of scientific fields we are familiar with nowadays, but also to the many areas of human life that are not within the scope of science. Apparently, there are many more perspectives than what science tries to capture with its barometers, thermometers, and spectrometers. Science may have its own window on the world, its own scientific point of view, but there are many other windows, views, vistas, or whatever you wish to call them. Reality is like a jewel with many facets that can be looked at from various angles, with different eyes. The late British philosopher Gilbert Ryle phrased this idea in his own terminology: "the nuclear physicist, the theologian, the historian, the lyric poet and the man in the street produce very different, yet compatible and even complementary pictures of one and the same 'world.'"[17]

Just like the "physical eye" sees colors in nature, so the "artistic eye" sees beauty in nature, the "rational eye" sees truths and untruths, the "moral eye" sees rights and wrongs, and the "religious eye" sees a divine dimension in life. All these "eyes" are in search of reality, but each one "sees" a different aspect of it—and therefore sees different "facts." As a consequence, there is so much more to life than what meets the scientific eye; there are many more ways of looking at things.

It should not surprise us then that there are also many kinds of blindness. Not only can one be physically blind, but also morally blind (those who sometimes fail to see what is right and what is wrong), or rationally blind (those who sometimes fail to see what is true and what is false), or even spiritually blind (those who cannot see beyond what is physical and material). We all have our blind spots and weak spots, but we could at least try to compensate for those.

The same holds for science. There is so much that science is "blind" for—at least a vast world of invisible entities. Science may be everywhere, but science is certainly not all there is. Science may be growing, but scientific expansion does not necessarily mean that other areas are shrinking. When it grows, science does not gain territory at all, but it just learns more and more details about its own fixed territory—which is the domain of what can be counted and measured. The rest is not part of its territory but was given away for other "authorities" to manage.

---

17. Gilbert Ryle, "The World of Science and the Everyday World," *Dilemmas* (Cambridge University Press, 1960), 68 - 81.

Yet, some scientists keep pursuing and promoting their superiority. Some state, for example, that physics always has the last word in observation because the observers themselves are physical. But why not say then that psychology always has the last word because these observers make for very interesting psychological objects as well. Either statement is philosophical nonsense; observers are neither physical nor psychological or whatever, but they can indeed be studied from a physical, biological, psychological, or even statistical viewpoint—which is a totally different issue. You cannot look at the world from one specific perspective and then claim this makes you see everything—that is, literally, all there is.

On the contrary, one and the same event can be looked at from quite different angles or perspectives, creating different frames of reference, and therefore different facts. But always be aware that it remains the very same world that we are talking about—no matter whether we look at it from a religious or a scientific viewpoint. The late Harvard biologist Stephen Jay Gould put it this way: "Science can work only with naturalistic explanations; it can neither affirm nor deny other types of actors (like God) in other spheres (the moral realm, for example)."[18]

Yet, this sound distinction does not deter some scientists from still claiming that they are in search of the "Great Theory of Everything" or, even worse, think they have already found it. Imagine such a theory—one single theory that would explain everything in this world! The atomic theory once had this disreputable aura—not only declaring that atoms are everywhere, but also that they are all there is, which is called materialism.[19] Later on, it became an important philosophical doctrine to decree that all living objects, including human beings, are only and merely machine-like automata, which is called mechanicism.

Then the philosopher Herbert Spencer came up with the proclamation that literally everything in life is based on natural selection, including our social relationships, motives, desires, beliefs, and morals. Spencer thus exported the theory of evolution from the biological realm to the realm of

18. Stephen Jay Gould, "Impeaching a Self-Appointed Judge," *Scientific American*, 267 (July 1992): 118–21.

19. Stephen M. Barr, *Modern Physics and Ancient Faith*, (University of Notre Dame Press, 2006).

sociology, psychology, and ethics.[20] Apparently, "survival of the fittest" had become the grand new "theory of everything" for Spencer and his allies. However, I have bad news for them—if rationality, morality, and even religion are the mere product of natural selection, so is science.

By the way, do not confuse a "grand theory of everything" with the "grand unification" or "grand unified theory" that physicists are dreaming of. These physicists are merely in search of a *physical* "umbrella" theory that would unify the three non-gravitational forces[21] as aspects of one single underlying force. No matter what the outcome will be, a search for a physical "umbrella" theory seems to be a very legitimate undertaking. But again, such a "grand unified theory of *physics*" is not the same as a "grand theory of *everything*." Even if such a physical "umbrella" theory would ever be found, it would only explain all *physical* facts in this universe, but it would never amount to a super-theory explaining everything else in this world as well—it would not qualify as a "Great Theory of Everything in this Universe," so to speak.

Even if physics could ever explain why the universe is the way it is, including its physical constants, we would still be left with the meta-physical question as to where the laws come from that explain everything that exists. Besides, we should always realize that nothing is final in science; all current scientific theories are the best we have right now, but only time will tell if they will hold out.

In spite of this caveat, it remains a timeless temptation for scientists to claim that the unknown has been reduced to almost nothing. I have bad news for them: The magnitude of the unknown is, well… unknown! The unknown is still in darkness, until we have the right searchlights or hypotheses to help us see in the darkness. The late American physicist John A. Wheeler, who coined the term "black hole," said, "We live on an island surrounded by a sea of ignorance. As our island of knowledge grows, so does the shore of our ignorance."[22]

---

20. Spencer wrote four books to promote his new "religion" based on his slogan "survival of the fittest": *Principles of Biology* (1864-67), *Principles of Psychology* (1870-72), *Principles of Sociology* (1876-96), and *Principles of Ethics* (1892-93).

21. Namely the electromagnetic, weak, and strong interactions.

22. Quoted in Clifford A. Pickover, *Wonders of Numbers* (New York: Oxford University Press, 2000), 195.

You may agree with what has been said so far, but yet have one important restriction: You do agree, but only as long as those different windows, aspects, perspectives, contexts, or whatever you like to call them, are based on science and have a scientific seal of approval. If you are of that opinion, then you have actually taken on the ideology of *scientism*—which is a dogmatic "creed" stating that science provides the only valid way of finding truth, thus eliminating everything that cannot be counted and measured.

Supporters of scientism claim that "the real world" is only a world of quantified material entities. They pretend that all our questions have a scientific answer phrased in terms of particles, quantities, and equations. Their claim is that there is no other point of view. Their trust is entirely in science, without trusting anything else, since they only acknowledge one territory—the territory of science. They believe there is no corner of the universe, no dimension of reality, no feature of human existence beyond its reach. In short, they have a dogmatic belief in the omni-competence of science.

In fact, however, they act like the drunken man who thinks that his lost car keys must be near the lamp post, because that is the only place where he can see in the light. To best characterize this attitude I like to borrow an image from the late psychologist Abraham Maslow: If you only have a hammer, every problem begins to look like a nail.[23] So I suggest not idolizing your "scientific hammer," because not everything is a "nail." Even if we were to agree that the scientific method gives us better testable results than other sources of knowledge, this would not entitle us to claim that only the scientific method gives us genuine knowledge of reality. I do admit, though, that if science does not go to its limits, it is a failure, but I must also add this warning: As soon as science oversteps its limits, it becomes arrogant—a know-it-all.

Curiously enough, those who defend scientism maintain that science is the only way of achieving valid knowledge, but they seem to be unaware of the fact that scientism itself does not follow its own rule. How could science ever prove all by itself that science is the only way of finding truth? There is no experiment that could do the trick! Science cannot pull itself up by its own bootstraps any more than an electric generator could run

---

23. Abraham H. Maslow, *The Psychology of Science* (HarperCollins, 1966), 15.

on its own power. So the truth of the statement "no statements are true unless they can be proven scientifically" cannot itself be proven scientifically. Scientism is in essence an im-material and self-refuting claim about the material world—if we consider it true, then it becomes false.

Scientism is also a baseless, unscientific claim that can only be made from outside the scientific realm, thus grossly overstepping the boundaries of science. It steps outside science to claim that there is nothing outside science and that there is no other point of view. That does not seem to be a very scientific move! It declares that everything that cannot be counted does not count. Why would such a declaration count given the fact that it cannot be counted? The late UCB philosopher of science Paul Feyerabend comes to the right conclusion when he says that "science should be taught as one view among many and not as the one and only road to truth and reality."[24]

Then there is another problem with scientism. It is an ideology that has us shackled in a physical, material world. However, making the claim that there is only physical matter implies that this very claim does not and cannot exist because claims are essentially *not* physical. Therefore, if we think that such a non-physical claim does exist, there must be more than physical matter in this universe. When we deny the existence of things immaterial, we also deny the existence of our own immaterial denial—as well as all our scientific claims. How could there be truth to what science claims if everything were merely material? Scientism is in fact self-defeating; it declares everything outside science as a despicable form of metaphysics, in defiance of the fact that all those who reject metaphysics are in fact committing their own version of metaphysics. Metaphysics may be a "dirty word" to some, but we all are surrounded by it—like it or not.

In short, there is so much more to life than science, for there is so much more that "counts" in life than can be counted. Science just can not account for all that needs to be accounted for. We should never forget that there are so many aspects of life that are off-limits for science. The astonishing successes of science have not been gained by answering every kind of question but precisely by refusing to do so.

As a result, scientific knowledge is not a superior form of knowledge; it may be more easily testable than other kinds, but it is also very restricted

24. Paul Feyerabend, *Against Method: Outline of an Anarchistic Theory of Knowledge* (New York: Verso Books, 1975), viii.

and therefore requires additional forms of knowledge. Mathematical knowledge, for instance, is the most secure form of knowledge but it is basically about nothing. Consider this analogy: A metal detector is a perfect tool to locate metals, but there is more to this world than metals. That is exactly where scientism goes wrong: Instead of letting reality determine which techniques are appropriate for which parts of reality, scientism lets its favorite technique dictate what is "real" in life—in denial of the fact that science has purchased success at the cost of limiting its ambition.

To conclude, I would say we have found a rather large opening in the seemingly closed fence around the scientific territory. Scientists may have closed *their* world, but that does not mean they also have closed *the* world. It always baffles me how science can enlighten and blind us at the same time by uncovering the material world while, at the same time, obscuring the immaterial world behind its undertakings.

## What Has Happened to Our Free Will?

If determinism were an all-pervading, inescapable phenomenon in this world—as science seems to suggest—such a view would create trouble for all those who claim that humans have a free will. And yet, having a free will is essential for us to be masters of our actions, not victims—by allowing us to make our own choices and decisions in life. Human freedom means that one is able to choose and act according to the "dictates" of one's own will—which is the freedom of self-determination. It is by our free will that we can and do shape our own lives.

Is there such a "thing" as free will? Unlike animals, human beings have this strong desire to become someone of their own making. To be sure, animals may have drives, impulses, instincts, emotions, and motives, but these are very different from intentions or reasons; not only are they mostly inborn, they also are always directly or indirectly related to sex or food. Even when it comes to food, humans often prefer to cook it first—an idea that animals have never come up with.

Besides, humans have the power of reason to guide them when emotions tend to take over. In addition, they have many other kinds of goals in life; not only do they live their lives based on role models, but also they steer their lives guided by reasons, purposes, plans, beliefs, values, hopes, dreams, and ideals—which shape them the way they are going to be. How different this is from animals. In the course of its life, an animal does not change much; it just looks and acts older and more worn-out. Humans, on the other hand, may have gone through dramatic changes in outlook on life, attitude, career, wisdom, faith, and beliefs—and hopefully for the better.

Nevertheless, the doctrine of complete determinism would not allow for such free decisions and choices, because everything in life is supposedly fully predetermined and preordained. But is determinism really an all-pervading phenomenon? Is our universe completely at the mercy of "law and order"? Is everything in this world really under the strict control of predetermined chains of causes and effects? Does the rigid "law-and-order" world that we live in and that science tells us about really nullify one of the privileges we thought we had, human freedom?

I have my strong doubts whether that is so! First of all, we discussed already that determinism in its all-pervasive version is self-defeating—if we consider it true, then it becomes false. When I say that all human beings are liars, then the very statement I am making as a human being must be a lie too. Similarly, when determinism wants to include everything in the universe, it must include the doctrine of determinism as well, which makes this very doctrine also the result of an inescapable chain of causes and effects.

In other words, trying to convince others of the doctrine of complete determinism would be a useless enterprise, since convictions are supposedly pre-determined as well. We cannot freely claim that free claims do *not* exist; and we certainly could not convince or persuade anyone else that this is true. Believing in complete determinism takes us into a vicious circle: If there is only room for determined things in determinism, then the *claim* of determinism must be one of those determined things—so it is not something I can freely claim. The claim of "hard" determinism just destroys itself.

Of course, we could safeguard the doctrine of complete determinism by claiming that it cannot refer to itself. But that is a desperate strategy, for if we let in one statement that is not pre-determined, many others may follow. So I suggest that it would be a safer strategy to limit the scope of determinism to smaller domains—for instance, the domain of science, and more in particular the various models science uses. To expand a limited domain to "everything there is" would be a lethal maneuver: If all my beliefs were fully determined by the motions of atoms in my brain, I would have no reason to suppose that my beliefs are true ... and hence I would have no reason for believing that my beliefs are dictated by atoms.[25] This may sound like a heavy philosophical statement, but I think it is worth pondering it in order to digest its implications.

But there is more. As we discovered earlier, laws of nature only apply under the conditions outlined in a model. Because of this, any specific case of determinism is a feature of the model; it cannot be applied as such outside the model. Models always select, so there is so much that they neglect. Put differently, each case of determinism may work fine within the framework of a particular scientific model, but it cannot be extrapolated and transposed to all other domains outside that framework—which certainly takes the edge off of it. Determinism is not so much a monopolistic world-view as it is a smart methodology for those working with scientific models.

Let us distinguish from now on between methodological determinism and ontological (or metaphysical) determinism. Methodological determinism is a harmless but powerful *technique* in science; it applies the law of cause-and-effect to a simplified version of the object under investigation. It is a handy tool to study things in the "test-tube-like" shelter of a model, which acts as a simplified replica of the original—that is, with a limited scope and under specific boundary conditions. But once we go outside the model and export its findings unchanged, methodological determinism becomes ontological determinism—a "hard-core" kind of determinism, actually a world-view that claims that what holds for the model also holds for the entire world.

Yet, "complexity simplified" remains the hallmark of science, because science is indeed, in the words of the Nobel Laureate Sir Peter Medawar, "the art of the soluble."[26] "Good" scientists are those able to demarcate their area of investigation by limiting themselves to factors that are relevant to

---

25. This is basically a paraphrase of what the biologist J.B.S. Haldane said (and C. S. Lewis repeated): "If my mental processes are determined wholly by the motions of atoms in my brain, I have no reason to suppose that my beliefs are true ... and hence I have no reason for supposing my brain to be composed of atoms." (*Possible Worlds and Other Essays,* Harper and Brothers, 1928, 209.)

26. P. B. Medawar, *The Art of the Soluble* (New York: Barnes and Noble, 1967).

what they are studying, and keeping strict control over factors that might interfere with their search. Let me use the following example to illustrate this.[27] When a computer calculates your income tax, the outcome can be determined—or predetermined if you will—by at least three completely different models. The mathematical model determines the outcome based on certain laws of income-tax arithmetic. The electronic model determines the outcome according to certain laws of physics operating inside the computer. And the accountant model determines the outcome by feeding the computer certain data and leaving other ones out. All of these are forms of "harnessed" determinism; in each case the outcome is predictable within its own simplified and limited setting. Each model selects its own setting and neglects everything else.

Interestingly enough, determinism is often associated with another technique: reductionism. Reductionism is an added useful tool in science, because it breaks the object under investigation down into a set of smaller and smaller components which can be easier studied in isolation than within the complex setting of the original. Reductionism tries to explain things by analyzing them into smaller and smaller parts—which is the "piecemeal" approach scientists excel in. The reductionists' creed says: After dissecting them into smaller pieces, things are much simpler than they appear. However, be careful not to mistake "simpler" for "easier to understand"; it is very doubtful whether an atom has a simpler "nature" than a molecule; smaller parts are not always really simpler, as quantum physics has shown us.

Nevertheless, most sciences have been very successful by creating a test-tube-like shelter in a laboratory, removed from the complexity of the real world. It is so much easier to study something *in vitro* than *in vivo*. It is in the test-tube-like setting of a model that all boundary conditions can be isolated, while other interfering factors can be kept under control. Immunology, for instance, becomes very complicated when we export it from the model 'in vitro' to the 'in vivo' world of organisms.

I would call this the *methodological* version of reductionism. But again there is also an *ontological* (or metaphysical) version that claims that the complex original *is* in fact the same as its much simpler parts and that the

27. I borrowed this example from the late physicist and philosopher Donald M. MacKay: *Science and the Quest for Meaning*, (Grand Rapids: Eerdmans Publishing, 1982), 24-25.

original is really "nothing more than" the sum total of its composing elements. In this view, organisms are just a collection of cells, cells are merely a collection of molecules, molecules are just a collection of atoms, and so on. This boils down again to confusing the model with the real world it is a simplified replica of.

Take the case of a gene-pool model. It simulates how selection in the gene-pool may change allele frequencies by selectively promoting certain alleles, the variants of a gene. However, if we mistake the model for what it represents, the model may give us the impression that the units of selection are in fact genes and their alleles, whereas "in the real world" they are not; in the real world, the organism is the unit of selection. The late Harvard biologist Stephen Jay Gould rightly noted that "selection simply cannot see genes,"[28] because alleles that do not come to expression in the organism cannot be subject to selection—selection just cannot "see" them. In the model, genes are the units of selection, but not so in the real world. Does this make the model useless? Far from that, but to identify the real world with one of its models would amount to ontological reductionism: A limited technique of simplifying what is complex has become an all-pervading doctrine or ideology of declaring all complexity as a matter of mere simplicity!

Let me also use the example of the human genome—which is a model in itself, by the way—to further explain my point. When we dissect the genome into individual genes, we can develop a model of genetic determinism that is very powerful (to a certain degree). However, a genome is more than a cluster of genes. What then are the limitations of the gene model?

First of all, some DNA sections, so-called *introns*, get initially transcribed into mRNA but are then removed from the end-product by splicing; it is due to alternative splicing that a single gene may code for several different proteins. This would partially explain why the number of genes can be much lower than we had initially expected.

Second, it turned out, as we discussed earlier, that protein-coding regions of genes can be interrupted by DNA segments that play more of a regulatory role by producing activator and repressor proteins that either activate or repress the activity of a "regular" gene.

---

28. Stephen J. Gould, "Caring Groups and Selfish Genes," *Natural History*, 86, 12 (1977), 20–24.

Third, genes can even be overlapping; about 9% of human protein-coding genes overlap another such gene. Sometimes the overlaps are partial, but in other cases small protein-coding genes are fully embedded within much larger genes (e. g. the blood clotting factor VIII)—so they are genes-within-genes. In this way, a nucleotide sequence may make a contribution to the function of one or more gene products. In other cases, some DNA sequences do double duty, encoding one protein when read along one strand, and a second protein when read in the opposite direction along the other strand. All these case may very well represent a hidden source of complexity to modulate gene expression.

And then there has been at least one more important development. Genes may be separated by long stretches of DNA that do not seem to be doing much at all—that is why they are often called "junk DNA," in spite of the fact that "non-coding," "neutral," or "silent" DNA would be a much safer term. Some of this "non-coding" DNA is repetitive DNA, often replicated from regular, coding DNA, and perhaps a rich source from which potentially useful new genes can emerge during evolution.

Speaking of simple genetic inheritance, the story has gotten more and more complicated. As a result, DNA looks more like an archive with instructions, but the issue of *what* to use from this archive *when*, *where*, and *how* is regulated in a much more perplexing way. The determinism of the gene model may no longer hold when applied to a wider setting. So we are never completely at the mercy of our DNA.

Let us place this in a more philosophical context. The late Hungarian-British physical-chemist and philosopher of science Michael Polanyi became famous—some would say infamous—for stating that a machine based on the laws of physics is not explicable by the laws of physics alone, because the structure of a machine is what he calls a "boundary condition" extraneous to the process it delineates.[29] Man-made machines, for instance, are only functional and useful because the laws of nature have been harnessed within certain boundary conditions or restraints.

Why wouldn't the same hold also for the "living machines" found in nature? The simple *parts* of an organism can only function within the boundary conditions of the complex *whole*. More in particular, the genes of

29. Michael Polanyi, "Life's Irreducible Structure," *Science* 160 (June 1968): 1308–1312. Or: *The Tacit Dimension* (New York: Double Day, 1966), 38.

the human genome have been harnessed within the boundary conditions of the entire genome—or actually of the entire organism.[30] There is some sort of hierarchy here: DNA atoms are harnessed within the structure of DNA; DNA molecules are harnessed within the structure of cells; and cells are harnessed within the structure of an organism. Whereas the genome is just a cluster of genes, the genotype is the blueprint of a genome plus the instructions for processing the blueprint.

Polanyi teaches us that principles and processes of a "higher" structure cannot be derived from principles and processes of a "lower" structure alone, just as it is impossible to derive a grammar from a vocabulary, or a vocabulary from phonetics. This makes some remark that the statement "the secret of DNA is life" is at least as compelling as the more popular slogan "the secret of life is DNA."[31] Think of viruses, which are essentially pure DNA or RNA; their DNA or RNA cannot do anything until they penetrate, like a Trojan horse, the interior of a "living" cell. They need the boundary conditions of a cell in order to fully operate.

This discussion may help us understand better why—once we step outside the model—determinism has lost its hard core. Let me explain this further with another example, this time taken from outside the scientific realm. When we watch a game on the golf course or on the pool table, we see balls following precisely determined courses of cause and effect; they follow physical laws and are subject to well-known rules. Yet there is one element that does not seem to fit in this pre-determined picture, in this cascade of causes and effects—the players of the game themselves. They may work with a physical model in their minds but they themselves are not part of that model. That is why the direction of something like a billiard ball on the pool table or a golf ball on the golf course is not only ruled by physical laws but also by human intentions—by players who have a certain goal in mind. And those very intentions do have consequences; they can become causes that are not part of the physical model but may have physical or non-physical effects on their own. The mind can obviously create its own non-physical causes in the midst of physical causes.

---

30. I am not going into the discussion here as to how specific DNA sequences came along. An account of how a system works is different from how it came into existence.

31. B. Commoner, "Roles of Deoxyribonucleic Acid in Inheritance," *Nature*, 202 (1964): 960-968.

Although we have a cascade of physical causes and effects in all these games, there is much more going on in each process—these players have a very specific intention in mind, which eludes and transcends the laws of science. Do they go *against* the laws of nature? Of course, they do not, and cannot; but they do go *beyond* those laws. People who are unable to look beyond those physical laws and causes are completely missing out on what the game is all about. These players somehow fall outside the realm of the model of physics; they themselves can steer the course of the laws of nature from "outside the model."

In other words, even in a world ruled by the law of cause-and-effect, there is also our own ability—at a "higher" level, so to speak—to be the cause of events all by ourselves. When actors act, their acts become causes with an effect. By knowing how things are determined by the laws of nature, human beings are able to construct highly effective and successful designs on earth—in architecture, engineering, and what have you. In doing so, they have the capacity to channel those "rigid" laws within a larger setting. Although chains of cause and effect appear to be very fixed and rigid, they can obviously be used and channeled within more extensive settings or designs. So human freedom still stands tall!

Freedom of self-determination does not mean, of course, that we can do whatever we choose! When calling this universe orderly, I mean that it is law-abiding—but not necessarily pre-ordained. So the freedom of self-determination does not let us do whatever we want to do, but it leaves us a series of options, curbed by a set of lawful constraints such as those known from physics, biology, psychology, sociology, economics, or history. The more we are aware of our constraints, the more we can actually be free. Since freedom is not free, we have to fight for it: "Know Yourself," says an old inscription in Delphi. With the proper knowledge, we can take charge of our constraints so that we are no longer their victims, but rather their architects. That is basically what engineers do too when they construct new machines.

In this connection, I should perhaps mention at least one caveat. The world of scientific laws has at least one big unknown, and that is the fact that the laws of nature as we know them may not be the laws of nature as they really are. Since our rationality is fallible, our knowledge may be fallible. Our scientific knowledge is by its very nature provisional "until further notice," because our theories and hypotheses are always tentative. We come up with what we think is true, and then check in experiments if what we think is true conforms to the way reality actually is. Only then may inventions become discoveries. If confirmed, we wait for additional confirmation. If not, we try it again by coming up with another, hopefully better conjecture. Science is an ongoing process of sifting out true ideas from false ones. Because of this, sometimes it may just *look as if* things go against scientific laws, but they may not upon further investigation.

These considerations take us inevitably into the spiritual realm—which I like to call our fourth (or fifth) dimension. It is the realm where the free will dwells under the guidance of *rationality* and *morality*. This extra dimension is an intangible dimension, outside the scope of science, but nevertheless real and true. If it were not, all we experience would be an illusion of the brain—including our scientific reasoning.

The only way to make human consciousness, human reasoning, and human exploration real and true is to face the reality of this extra dimension. The intangible must be as real as the tangible. If we were just biological beings, then our mental activities would just be the effect of a neural mechanism, not its cause. They would be merely images, mirages, illusions—but nothing real and true. Only if we are more than something physical and biological, our mental activities can become real and true. Those, however, who have lost this awareness of an extra dimension, have lost the dimension of the invisible—even in science. In fact they have lost much more: the dimension of the supernatural, the dimension of the spiritual and divine, the dimension of rationality and morality.

We are entering here the world of soul, spirit, and mind. If you want to distinguish them further, I would say the mind is the intellectual part of the soul; yet both belong to our spiritual side. The mind is the power of the soul by which we know truth; the mind is the soul's "eye," its "light," so to speak.

Just as we have five senses for the material aspects of this world, so we also have at least two "senses" for the immaterial aspects: the *rational* sense of true and false as well as the *moral* sense of right and wrong. If we were only the fortuitous effects of physical causes, we would have no other rational and moral measures but ourselves. Denying the existence of immaterial activities in human beings is in itself an immaterial activity, and thus leads to contradiction—actually self-destruction. Therefore I would say we must be more than matter.

The sciences deal only with the physical senses. As the saying goes, do not touch anything in a physics lab, do not taste anything in a chemistry lab, and do not smell anything in a biology lab—and do not believe everything you hear in a philosophy department, I might add. However, my physical senses tell me also that my body will die certainly and inevitably. But fortunately, my other "senses," my mind and my reason, tell me I am more than a material machine.

Whereas the body and its brain are tangible, the soul and its mind are intangible. Through our minds, we know if what we know and experience is accurate and real, and we recognize that we can discover, through observation and experimentation, how the material world works. We know with virtual certainty that logic, reason, science, and mathematics are reliable ways of learning the truth. So we know that we exist and that we can make choices and decisions in life guided by our rational and moral "senses." And decisions do have consequences, as they become their own causes with effects.

Let me use the following example to explain how the tangible and intangible differ. *Blinking* is something we can capture with a camera, but *winking* is of a rather different nature; it is not just a single blink with one eye (unless you suffer from a winking "tic"), but there is a very particular intention or purpose behind winking. What does a "wink" add to a "blink" then? The answer is straightforward: My wink tells you to read my *mind* as kidding. There is something behind a wink that a camera cannot capture.

Would you still maintain that winking is less "real" than blinking? To be sure, winking at someone is definitely a physical process, as much as blinking is, but it is also very different from blinking an eye, as it surpasses a series of physical causes. In contrast to an eye blinking there is something more behind an eye winking—call it a "mind" or a "soul" that decides to wink. The late Nobel-laureate and physicist Arthur Compton, who discovered the 'Compton effect,'[32] put it this way: "If the laws of physics ever should come to contradict my conviction that I can move my little finger at will, then all

32. The *Compton effect* is the increase in the wavelengths of X rays and gamma rays when they collide with and are scattered from loosely bound electrons in matter.

the laws of physics should be revised and reformulated."[33] In other words, decisions, although immaterial, do have consequences, even material ones.

One time, the late neurosurgeon Wilder Penfield asked his patient to try and resist the movement of the patient's left arm that he was about to make move by stimulating the motor cortex in the right hemisphere of the brain. The patient grabbed his left arm with his right hand, attempting to restrict the movement that was to be induced by a surgical stimulation of the right brain. As Penfield said, "Behind the brain action of one hemisphere was the patient's mind. Behind the action of the other hemisphere was the electrode."[34] That is the enigma of the mind we are heading for.

## We Are Rational and Moral Beings

At this point, I would assume, we begin to have a sneaky suspicion that we must be more than what science tells us that we are. If you agree with me, we can perhaps set a step farther and enter another stage in our thinking: We are beings with a free will—free to make decisions, rational or irrational, moral or immoral decisions. In other words, we are endowed with the faculties of rationality and morality. Our lives are supposed to be under the guidance of rationality and morality. What does this mean?

As *rational* beings, endowed with the capacity of rationality, we are in pursuit of what is *true* (versus false). Rationality is our capacity for abstract thinking and having reasons for our thoughts, thus giving us access to the "unseen" world of thoughts, laws, and truths—allowing us to gain knowledge and thus be masters of our own actions. The power of reason is an important compass when emotions tend to take over. Rationality allows us to gain knowledge about the world through the power of abstract concepts and mental reasoning.

Reasoning leads us from one idea to a related idea; it is a matter of pondering realities beyond that which we experience through our senses, thus allowing us to transcend the current situation with the mental power

---

33. Arthur Compton, *Man's Destiny in Eternity. A Book from a Symposium. The Garvin Lectures*, (Boston: Beacon Press, 1949).

34. Wilder Penfield in the "Control of the Mind," a Symposium held at the University of California Medical Centre, San Francisco, 1961, quoted by A. Koestler, *Ghost In the Machine* (London: Hutchinson, 1967), 203f.

of abstract concepts and mental reasoning. Philosophical giants such as Aristotle and Thomas Aquinas would put it this way: All we know about the world comes through our physical senses but this is then processed by the immaterial intellect that extracts from sensory experiences that which is *intelligible*.[35] It is rationality that makes the world intelligible and understandable; it gives us the power to comprehend the universe.

Without the human mind, without its intellect and rationality, there would not be any science—or worse, there would not even be any facts, for facts are mental interpretations of this world, made through our intellect, as we discussed earlier. Animals certainly live in a world of events, but humans also inhabit a world of facts—which are mental interpretations of those events. The world of animals is only populated with what their physical senses can capture, whereas capturing facts requires a spiritual mind, endowed with something like a sixth sense. That is where rationality comes in. It is in search of what is true, so as to avoid rational mistakes as much as possible. It is essential to all of us that we can discern if human beings are telling the truth or not.

As *moral* beings, endowed with the capacity of morality, we are in pursuit of what is *right* (versus wrong). Not everything that is thinkable or possible or reasonable is also permissible. Morality is about our rights and obligations, about what actions others owe us and what actions we owe to others, as part of the "common good." Duties and rights have a natural reciprocity: The duty of self-preservation is also the right of self-preservation; the duty to seek the truth is also the right to seek it; the duty to work for justice is also the right to work for it.

So morality is not about what the world *is* like, but about what the world should be like; it is not a matter of description but prescription. "Racial equality," for instance, is not a descriptive but prescriptive term; races are not equal in biological characteristics but they do have the same dignity and rights. Morality tells us what *ought* to be done in life—no matter what, whether we like it or not, whether we feel it or not, whether we

35. The Scholastics used to say that there is nothing in the mind that was not first in the senses. What they meant is that what is immediately sensed—"qualities" or "impressions" in Hume's terminology such as colors, sounds, and odors—are only the media through which reality is discerned and understood, but we are not confined to those sensations.

want it or not, whether others enforce it or not. It tells us what ought to be done—by us, as a duty, and towards us, as a right—otherwise a moral mistake would be made.

This having been said, morality is not a matter of political correctness but of moral correctness. It makes us search for what is intrinsically right and what is intrinsically wrong. Whereas our bodily movements are subject to physical constraints, our actions are subject to moral ones. It is essential to all of us that we can discern if human beings are doing the right thing or not.

Rationality and morality seem to set us apart as human beings, for they are not coded for in our DNA and therefore do not come from the animal world. Let me demonstrate this for rationality first. Rationality is not a matter of intelligence but of intellect; whereas intelligence can be graded on an IQ scale, intellect cannot. For those who question my distinction between intelligence and intellect, I have added the following two, rather technical, paragraphs—skip them if you want to.

Intelligence works only with perception of sense-data. Many animals show some form of intelligence in their behavior, because intelligence is a brain feature and as such an important tool in survival. Animals can process images more or less intelligently. They show various forms of intelligence: We find spatial intelligence in pigeons and bats, social intelligence in wolves and monkeys, formal intelligence in apes and dolphins, practical intelligence in rats and ravens, to name just a few. Intelligence is a matter of processing sense-data—something even a robot can do by "cleverly" processing sounds, images, and the like.

Intellect is very different from this. Like intelligence, intellect also uses sense-data such as images and sounds, but unlike intelligence, it changes perception into cognition by using concepts and reasoning, and thus making sensorial experiences *intelligible* for the human mind. A concept may be as simple as a "circle" or as complex as a "gene," but a concept definitely goes beyond what the senses provide. Concepts are very different from images. Images are by nature ambiguous, open to various interpretations; we need concepts to interpret them. We do not see genes but have come to hypothesize them in a concept. We do not even see circles, for a "circle" is a highly abstract, idealized concept (with a radius and diameter). Once these concepts have been established and mastered, we have become "regular observers" of "circles" and "genes." But again, they are not images but concepts.

Mental concepts transform "things" of the world into "objects" of knowledge; they change experiences into observations, thus enabling humans to see with their "mental eyes" what no physical eyes could ever see before. To be sure, all we know about the world does come through our physical senses but this is then processed by the immaterial intellect that extracts from sensory experiences that which is intelligible with the aid of concepts. Obviously the intellect may assist human intelligence, but intelligence can also work on its own, as it does in animals and even robots. But the latter do not have intellect, because they lack mental concepts and conscious reasoning. You can have more or less intelligence, but you cannot have more or less intellect—humans have it, animals do not.

Back to rationality, rationality is an intellectual "tool." It gives us cognition in addition to perception; it gives us the mental power of abstract concepts and logical reasoning; it enables us to ask questions, to formulate concepts, to frame propositions, and to draw conclusions. The rationality of our intellect gives us the capacity for abstract thinking and having reasons for our thoughts, thus giving us access to the "unseen" world of thoughts, laws, and truths—allowing us to become masters of our own actions and experts in science. In short, reasoning is pondering realities beyond what we experience through our senses.

Animals, on the other hand, are not rational—they may be more or less intelligent, but they are not rational beings. Pets can even be smarter than their owners when they play a whole repertoire of tricks on their owner's emotions—but that is a matter of intelligence, not intellect. Animals do have the capacity to sense, imagine, and remember things, but they lack understanding in the sense of interpreting images, asking questions, formulating concepts, framing propositions, and drawing conclusions. They show no signs of abstract reasoning or having reasons for their "thoughts" (if they have any); they do not think in terms of true and false; they do not think in terms of cause-and-effect with "if-and-only-if" statements; they are "moved" by motives, drives, instincts, emotions, and training—but not by reasons or mental concepts.

Since we are masters of anthropomorphism, we tend to think that animals must be like humans, even with regard to rationality. Some consider it politically incorrect to refer to them as "pets"—they are to be referred to as "domestic companions." But what a disparity there is between them and us! Animals live their lives entirely in the present, without having any thoughts about the past or the future—perhaps memories, but not thoughts. If animals have a pedigree, it is thanks to their owners; if they have birthdays, wish lists, appointments, or schedules, it is because their owners create those; and if they have graves, those were dug by their owners as well. Only humans have the capacity to even discuss the respective identities of humans and animals.

Only humans can ponder realities beyond that which is experienced through the senses. Only humans are conscious of time; they can study the past, recognize the present, and anticipate the future; they even desire to transcend time, thinking about living forever. Only humans wonder "what caused or will cause what and why?" Only human beings have inquisitive minds asking questions such as "Where do we come from?" and "Why are we here?" Only humans have the capacity to be scholars and scientists; we can study animals but animals cannot study us—they can watch us but not study us. Human beings are always in search of some kind of worldview or explanation of life—which certainly goes far beyond their need for food. In short, human beings are questioning beings; they are driven by rationality, which gives them the capacity to make rational decisions (without any guarantee, of course, that those decisions are always rational).

The cosmologist Stephen Hawking usually has not much good to say about human beings: "We are just an advanced breed of monkeys on a minor planet of a very average star," but he had at least the sensibility to add, "But we can understand the Universe. That makes us something very special."[36] That was surely a great afterthought he had!

There is a similar story about morality. In the animal world, there is no morality. Only humans can curb their animal instincts and drives with morality. But since animals do not have any moral values, they follow whatever "pops up" in their brains. The relationship between predator and prey, for example, has nothing to do with morality; if predators really had a conscience guided by morality, their lives would be pretty harsh. Dogs may act as if they are "caring," but they just follow their instinct, not some moral

36. Stephen Hawking in an interview with *Der Spiegel* on October 17, 1988.

code; dogs happen to have such an instinct, whereas cats lack it, since it is not in their genes.

As a consequence, animals never do awful things out of meanness or cruelty, for the simple reason that they have no morality—and thus no cruelty or meanness. But humans definitely do have the capacity of performing real atrocities. On the other hand, if animals do seem to do awful things, it is only because we as human beings consider their actions "awful" according to our standards of morality. Yet, we will never arrange court sessions for grizzly bears that maul hikers, because we know bears are not morally responsible for their actions.

Since animals have no moral values, they have no duties, no responsibilities, and consequently no rights. If animals had rights, their fellow animals also would need to respect those. We, as masters of anthropomorphism, may *think* they have morality, but they do not. On the other hand, since we do have morality, we need to treat animals, God's other creatures, humanely and responsibly—not because animals deserve it, but because humans owe it to their Maker and to themselves, being stewards of what the Maker created. As Antoine de Saint-Exupéry wrote in *The Little Prince*, "You are responsible forever for that which you have tamed."

So the question emerges as to where rationality and morality come from then. Although they do not seem to come from the animal world, they could still be located in our—uniquely human—genes. I have my doubts, though, that this is the case. Let me explain this with regards to rationality first. If rationality were really a matter of genes, it would belong to the material world and would therefore be as fragile as the material world itself. It would be sitting on a swamp of molecules, unable to pull itself up by its bootstraps. If rationality were really a matter of DNA, it would lose all its power. Claims can be true or false, but molecules such as DNA cannot be true or false.

In order for us to make any rational claims, we need to validate our claims as being true, otherwise they are worth nothing. If Watson and Crick were nothing but DNA, then Watson and Crick's theories about DNA must be as fragile as their DNA. That would be detrimental for their claims. If we were nothing but DNA, this very statement that we are making here would not be worth more than its molecular origin, and neither would we ourselves who are making such a statement. Claims like these just defeat and destroy themselves. They cut off the very branch that the person who makes such claims is—or actually was—sitting on.

If we want to accept the reliability of our biological knowledge regarding DNA, we cannot conclude at the same time that all human knowledge is just a product of DNA. That would be "irrational" suicide! If rationality were the product of DNA, we definitely should be questioning the validity of our knowledge—which necessarily includes all the scientific knowledge we wish to claim. In short, we would have no *reason* to trust our own reasoning. That would be the end of anything we claim to be true—a thought that truly stops all thought.

We would run into similar problems when it comes to morality. If morality were really a matter of genes, we should first of all question why we would need some articulated moral laws and rules to reinforce what "by nature" we supposedly would or would not desire to do anyway. As it happens, there are only far too many people willing to break moral rules when they can get away with it. Are they really fighting their genes; or do they perhaps have the "wrong" alleles?[37]

Take, for instance, the moral responsibility parents feel towards their underage children. Is this a natural, "instinctive" responsibility that was promoted by natural selection because it presumably improves the offspring's reproductive capacity? Apparently it is not, given the fact that far too many parents try to ignore their so-called "natural" responsibility. Or take monogamy: If we were monogamous "by nature", we would not need moral laws to protect family life. In other words, we are rather dealing here with responsibility in a *moral* sense.

Responsibility is not a genetic, but moral phenomenon. Moral laws tell us to do what our genes do not make us do "by nature." In morality, we must decide between inclinations to do what is good and temptations to do what is bad. Morality is not under the rule of natural selection, because any moral decisions we make are not inborn, and therefore cannot be subject to natural selection. The philosopher Jean Jacques Rousseau may have promoted

---

37. E.O. Wilson, for instance, insists that human rules against incest have a genetic basis (M. Harris and E. O. Wilson, "Encounter," *The Sciences,* 18 (1978): 10-15, 27-28). His reasoning comes close to a circular argument: Cultural universality suggests a genetic basis; because there is a genetic basis, incest is a universal phenomenon in human cultures.

the idea of man's "innate goodness" and of the "noble savage" who becomes only corrupted by organized society, but we all know better—we are all part of one large dysfunctional family. When kids bully others, for instance, they should be held accountable and disciplined, regardless their age.

If that is not enough to defuse genetic claims on morality, I have to add one more reason as to why such attempts necessarily fail. If morality were really a matter of DNA—that is, with a material basis, but not an objective basis—we would not be morally obligated, but would only *feel* obligated; our genetic make-up would only have us *believe* that our moral obligations rest on an objective foundation—a collective illusion, so to speak, foisted on us by our genes. Again, such a foundation would be as fragile as the DNA material it is said to be made of. We would have no *right* to claim any moral rights. That would be the end of anything moral that we are obligated to do to do to others and that others are obligated to do to us.

Needless to say that rationality and morality are closely intertwined. Rationality gives us *reason* to defend what is morally right. And morality gives us the *duty* to pursue what is rationally true; it also gives us the *right* to defend what is rationally true. Because of this interaction, a flawed rationality has an impact on morality, and a defective morality affects rationality. In other words, rationality and morality were given to us not as qualifications we "own" already but as capacities we can and should use. If you ever wonder why there is so much irrationality and immorality in this world, think of this: Rationality and morality are not inborn, so they have to be taught, cultivated, and nurtured, otherwise they wither. That is why we have sometimes reason to ask people whether they are "in their right mind."

We could also word this in terms of "maturity." Usually grown-ups are more mature than youngsters, but do not be surprised when a ten-year-old shows more maturity then a fifty-year-old who never matured by lack of spiritual "nourishment." If you still believe maturity is "in your genes," be aware that parents and children not only share their genes but also their environment, even as early as in the womb. Kids do learn from their parents, and later on also from their peers. On the one hand, a good role model can be very contagious; on the other hand, one rotten apple can spoil the whole barrel.

The conclusion after all of this seems rather clear to me, and I hope to you too. With rationality being in pursuit of what is true and with morality in pursuit of what is right, we find ourselves in a non-material world where things are not large or small, light or heavy, hard or soft, but they are true or false, and right or wrong. Rationality strives for what is universally and objectively *true* in the same way as morality strives for what is universally and objectively *right*.

Rationality is in search of universal laws and objective truths in this universe which tell us the way it *is* in this world—no matter what, whether we like it or not, whether we feel it or not, whether we want it or not. In a similar way, morality is in search of universal obligations and objective values in this world; these tell us what we *ought* to do—no matter what, whether we like it or not, whether we feel it or not, whether we want it or not. Yet, morality cannot be without rationality; morality needs the power of argument, not the power of force. It may not be im-morality that is the biggest threat to morality in our society but the lack of rationality, transforming morality into an incoherent language, merely a matter of emotions.[38]

As C.S. Lewis put it, "The human mind has no more power of inventing a new value than of imagining a new primary color."[39] Truths and rights are not invented but given—they are nonnegotiable. Without rationality, we would have nothing to steer our knowledge in terms of true and false, and without morality, we would have nothing to evaluate our actions in terms of right and wrong. Ironically, those who are opposed to "imposing value judgments" on others are happy to hold everyone to their own value judgment which states that we are not supposed to impose any value judgments on anyone.

Hence, if we want to entertain the notions of rational reasons and moral rights, we must accept that there is an immaterial world of rationality and morality where objective and universal laws reside. It would be our task then to discover those laws. In the same way as humanity tries to decipher the laws of the universe through science, human beings should also be in pursuit of the moral laws that are "written" in their hearts and steer their conscience;[40] they should not replace the moral law with idols of their

---

38. This is the main point of Alasdair MacIntyre's book *After Virtue* (University of Notre Dame Press, 2007).

39. C.S. Lewis, *The Abolition of Man* (New York, NY: Macmillan, 1973), 56–7.

40. Moral laws are often collectively referred to as "moral law" or "natural law."

own making. As human beings, we have been endowed with the rational power of our intellect and the moral power of our conscience.

We cannot give in to the absolute dictatorship of relativism, for relativists in morality—and in rationality as well—defy themselves when they make the absolute statement that everything is relative. Ironically, even moral relativists hold on to at least one moral absolute that says "Never disobey your own conscience." So we should ask them the question as to where the absolute authority of a human conscience comes from. If our conscience were merely a private issue that we form at our own discretion, it could never claim any absolute authority, for moral disagreements cannot be settled on the level of a person's private and personal conscience. If so, there would be a tie, for instance, between the conscience of a pro-choice pregnant mother and the conscience of her pro-life obstetrician. Each "private" conscience must ultimately be under the authority of a universal "natural law" of morality.

The conclusion of all of this seems to be clear to me: Without an immaterial world of reasons, I would have no *reason* to believe any of my rational beliefs. In the same vein, if there were no immaterial world of rights, I would have no *right* to claim any of my moral rights. To put it conversely: If I believe that rationality and morality are determined by DNA, I would have no reason to believe that my belief is true… and hence I would have no reason for supposing that rationality and morality are determined by DNA! In a world of molecules, there is talk of being small, heavy, strong, and what have you—but not of being true and false, or right and wrong.

One more remark: Since we are not rational and moral "by nature," we need articulated rational and moral rules to reinforce what "by nature" we would or would not desire to think or do anyway. That being said, I should add that rationality and morality are both fallible. Sometimes we say "I should have known better"—which refers to a cognitive failure—or "I should have acted better"—which points at a moral failure. In spite of such potential failures, we do have access to an immaterial world where natural laws and moral laws reside, so as to regulate what is true or false and right or wrong. Unlike animals, we realize that we know less than we could have known, and that we do less than we should have done. Since we have access to an immaterial world that we can compare ourselves with, we are familiar with the fact that we are always less than we should and could have been. Since we are not rational and moral by nature, we need to be *taught* to be rational and moral.

In this context, I like to quote Blaise Pascal, the inventor of mechanical calculators and an authority in statistics and hydrodynamics[41], when he said, "It is dangerous to show man in how many respects he resembles the lower animals, without pointing out his grandeur. It is also dangerous to direct his attention to his grandeur, without keeping him aware of his degradation."[42] Just as the brain can be affected by material forces such as alcohol and drugs, so the mind can be affected by spiritual forces that either inform or misinform us. Therein lies our power as well as weakness at the same time. In rationality, the power of reason can be overpowered by the force of deception; in morality, a clear sense of right and wrong can be overshadowed by the tyranny of unruly passions. When rationality and morality are not taught, cultured, and nurtured, they wither and get overpowered by irrationality and immorality.

No wonder then that people often like to excuse themselves for making the wrong decisions—no matter whether these are rational or moral decisions. A good candidate for this is genetic determinism—it offers a handy alibi for human responsibility.[43] Once we declare ourselves no longer responsible for our rational or moral decisions, we think we can accept what is *not* true or *not* right. If you wish to be "off the moral hook," you must find a way to defend the claim that you cannot be held responsible for your bad decisions. One way is to "geneticize" or "medicalize" flawed knowledge or immoral behavior: The victimizer is no longer a person but a disease or pathology caused by genes, hormones, or frontal lobes. All of a sudden, we find ourselves suffering from a "disease" supposedly beyond our control. Once we have made ourselves victims, we feel released from any responsibility, since victims are, by definition, not responsible for their conditions, but can point instead to something else as the culprit—genes, hormones, diseases, syndromes, and pathologies.

Unfortunately, this is exactly what is happening when people ignore

---

41. The physical unit of pressure—Pa—is named for Pascal.

42. Blaise Pascal, *Thoughts on Religion and Philosophy*, trans. Isaac Taylor (Edinburgh: Otto Schulze and Co., 1901), chap 1, VII.

43. Steven Rose, Leon J. Kamin and R. C. Lewontin, *Not in Our Genes: Biology, Ideology and Human Nature* (New York: Pantheon Books, 1985).

that their lives should be under the guidance of rationality and morality. When we say, for instance, that people deserve praise for their inventions, that they should be criticized for their wrong ideas, that they are held justly responsible for a crime, or that they deserve reward for a heroic act of self-sacrifice, we mean that they were the author and cause of what they did in such fashion that they had it in their power *not* to do what they did. Human freedom is the basis of praise or blame, merit or reproach. Apparently, human beings can certainly be causes on their own, apart from any physical causes around them.

I think my message is clear: If rationality and morality were mere products of DNA, we would never be able to declare our thoughts as being true or false and our moral judgments as being right or wrong—they would all be of the same caliber: matter. If they were just "matter," they would be worth nothing, or at least nothing more than the atoms DNA is made of. We would have no foot to stand on in order to decide between true and false reasoning or between right and wrong moral judgments. If our laws of nature and our moral laws were generated by DNA, they could never claim universal validity either. That seems to me a rather final and detrimental verdict on such a world-view.

What have we achieved so far? We may have discovered an opening in the scientific fence, but that in itself still leaves us alone in this universe—even more alone than ever, I would say, given our distinctive human capabilities of rationality and morality that separate us from the animal world. Those abilities might make us feel even more isolated from the rest of the world, unless...

# 3. Are We Alone?

Although science seems to suggest that we are alone and on our own in this universe, we have learned so far that the scientific story is quite limited and must leave many other aspects of this world necessarily and by its own nature untouched. What you think you know may not be so—not only in your science but also in your philosophy.

So the fact that everything may have started with the Big Bang does not mean that there is nothing more to the story. So perhaps there is actually *more* out there, somewhere "beyond" and "behind" the Big Bang. If that were true, then there might also be a Creator behind this universe. If that were true, then this universe is not such a God-forsaken place after all. If that were true, then we might not be so alone at all.

Augustine once said, "What place is there in me to which my God can come, what place that can receive the God who made heaven and earth? Does this mean, O Lord my God, that there is in me something fit to contain you?"[1] A more recent paraphrase was given by the famous philosopher, physicist, and mathematician Blaise Pascal: "What else does this craving, and this helplessness, proclaim but that there was once in man a true happiness, of which all that now remains is the empty print and trace? This he tries in vain to fill with everything around him [...] since this infinite abyss can be filled only with an infinite and immutable object; in other words by God himself."[2] Let us find out.

1. Augustine, *Confessions*, 1.1.1.

2. Blaise Pascal, *Pensées*, 10.148: 1.

## The Cause of All Causes

Is there a limit to what can be known? It depends. Some people are eager to create their own boundaries ahead of time as to what can be known. Members of philosophical schools such as positivism, logical-positivism, and language-analysis are masters at this; they define what is considered legitimate knowledge by making sure they exclude what they do not wish to be legitimate knowledge.

The famous Austrian-British philosopher Ludwig Wittgenstein started his career this way by claiming that we can only talk sensibly if our statements are either strictly empirical or strictly logical. In this view, one can legitimately say "the root of a plant absorbs nutrients" (which is an empirical statement), and one can also legitimately say "the root of four is two" (which is a logical statement), but the statement "the root of a plant is two" amounts to non-sense and is therefore rightly considered meaningless. In this kind of philosophy, it is clear ahead of time that any statement about the Great Unknown is anathema —an abuse of language. This made Wittgenstein exclaim, "Whereof one cannot speak, thereof one must be silent."[3] Otherwise we would be committing "dirty" metaphysics. Needless to say that it is hard to tell whether claims like these are either logical or empirical—or just meaningless.

It did not take Wittgenstein long, though, to see the limitation of his earlier views. He gradually began to make some pretty "senseless" statements, as per his old view. Here is just one of those: "To believe in a God means to see that the facts of the world are not the end of the matter."[4] Perhaps this may give us some awareness of the fact that rejecting "senseless" metaphysics can only be done on metaphysical grounds, for any rejection of metaphysics is based on a metaphysical viewpoint regarding what the world "really" is like. Metaphysics may be a "dirty word" to some, but we all are surrounded by it—like it or not. So let us go for some metaphysics then.

Earlier we discovered that determinism is based on the law of cause-and-effect. This law works great within the setting of scientific models, so

3. Ludwig Wittgenstein, *Tractatus Logico-Philosophicus*, translated by C. K. Ogden and D. F. Pears (London: Routledge, 1922), 7.

4. Ludwig Wittgenstein, *Notebooks*, 1914-1916, translated by G. E. M. Anscombe (Oxford: Basil Blackwell, 1961), 74e (8 July 1916).

we found out, because all scientists follow the rule—once formulated by the philosopher John Stuart Mill in 1872—that it is a law that every event depends on some law.[5] It was this rule that made Charles Darwin write in one of his letters, "[...] astronomers do not state that God directs the course of each comet and planet."[6] Darwin was right on target: Comets and planets just follow laws, and so must the rest of nature. The question remains of course what Mill's law is based on or comes from. We will tackle that question later on.

Just as one question leads to more and more questions, so does the search for a cause lead to searches for more and more causes. Little kids can drive you nuts by asking perplexing questions. They might ask you what makes the weather so hot, and you may answer that the sun causes this. This might evoke another question: What causes the sun to produce such heat? Even if you know how to answer that question, be prepared for the next one: And what then causes what you just told me? We are in for an endless back-and-forth. In a way, though, even adults never have outgrown that attitude. We still keep asking what causes this or that, and then go on asking what the cause of that cause is, and so on. In principle, there is no end to this series of questions.

If we want to follow this line of questioning, we could end up with the following dialog. Question: What caused our planet to be here? Answer: The Big Bang caused all of this. Question: What caused the Big Bang? Answer: Nothing. End of discussion? Well, there is something unsettling about the final answer, because "nothing" cannot cause anything except nothing—or, as they rightly say, "Nothing comes from nothing."

Most philosophers in the Western tradition would state that there is nothing that can explain its own existence. Instead they would say something along the following lines: "Nothing" just is without a reason why it is, whereas everything that is something has some adequate or sufficient reason why it is.[7] Let me repeat the question then: What caused the Big Bang? Instead of answering "nothing," we could have answered: God. But

---

5. John Stuart Mill, *A System of Logic* (Honolulu: University Press of the Pacific, 2002) chapter V, § 1.

6. In his correspondence with geologist Charles Lyell in 1861.

7. Leibniz coined the term "principle of sufficient reason": Nothing is without a ground or reason why it is (*Monadology*, 36).

that might trigger the next question: What caused God?—which is a question very unsettling as well, because we seem to be heading for an infinite regress of new questions, with no end in sight.[8]

Yet, I would argue that the correct answer can only be found in God. There must be some kind of explanation for this universe and for all that comes with it, including the Big Bang. Why? The answer that things are just the way they are is not a very satisfying response; our universe need not be the way it is, and it need not even exist. The Boston College philosopher Peter Kreeft sarcastically calls this the 'Pop Theory'—things just pop into existence for no reason at all.[9] Instead, he claims that we may never find the cause, but there *must* be a cause for everything that comes into existence. Rationality demands this (unless one decides to abandon our human capacity of rationality, but that is a position hard to defend on rational grounds).

In other words, our universe is neither necessary nor absolute, but finite and dependent instead; the more philosophical term is *contingent*. However, if there is no inherent necessity for the universe to exist then the universe is not self-explaining and therefore must find an explanation outside itself. Obviously, the universe cannot be grounded in something else that is also finite and not self-explaining for that would indeed lead to vicious infinite regress. Therefore, it can only derive from an unconditioned, infinite, and ultimate ground, which most people would call a Creator God.[10]

But would God not need a ground or cause as well? Would the explanation of a pre-existing designer not merely beg the question of who designed the designer? The answer is that only things that have a beginning need a

---

8. A series of propositions leads to infinite regress if the truth of the first one requires the support of a second one, the truth of the second one requires the support of a third one, and so on, and the truth of the next-to-last one requires the support of the last one, where the number of "last" approaches infinity.

9. Peter Kreeft, *Fundamentals of the Faith: Essays in Christian Apologetics* (San Francisco: Ignatius Press, 1988).

10. The German philosopher Gottfried Leibniz used to say that the *sufficient reason*, which needs not further reason, must be outside of this series of contingent things and is found in a substance which is a necessary being bearing the reason for its existence within itself. (*The Principles of Nature and Grace, Based on Reason*, 1714).

cause outside themselves. God, however, is eternal and absolute, and therefore does not have a beginning—he is not self-caused but un-caused. The universe, on the other hand, cannot be self-caused, because that would mean that the universe existed before it came into existence—which is a logical absurdity. As a consequence, there *must* be an Absolute Ground for all that happens to exist, even for the universe itself—unless you wish to abandon rationality. So it seems to me that we, as rational beings, are on our way towards the Great Unknown!

Let me try to clarify this with some more philosophical terms. Thomas Aquinas[11] calls God a "Primary Cause," and all the causes science deals with he would refer to as "secondary causes." The physical causality of science reigns "inside" the universe, linking causes together in a chain of secondary causes—the chains of causes and effects that we talked about in the previous chapters. God, on the other hand, reigns from "outside" the universe as a Primary Cause or First Cause, thus providing a "point of suspension" for the chain of secondary causes itself, so to speak.

I hate to use images to help us picture this, as analogies can be very misleading, but let me try this one. An old-fashioned grandfather's clock has an intricate mechanism of cogs, springs, and wheels, which interact with each other in the rigid way of cause-and-effect. The clock ticks due to the way these secondary causes affect each other. However, the clock will stop ticking if there is not someone—let us say, for a moment, some kind of a primary cause—who keeps periodically winding up the clock from outside the system. This person is not another cog, spring, or wheel inside the clock, but comes from outside the system. This is the dangerous image I dare to use in order to explain what God does to the universe. Let this imagery not mislead you, though: God is not the "watchmaker" who set the watch into motion and then abandoned it to itself.[12] I will clarify that later.

Then I have another word of caution. God as a Primary Cause is not a super-cause among other causes, but he is "above" and "beyond" all

---

11. A good introduction to Aquinas' philosophy is: Edward Feser, *Aquinas: A Beginner's Guide* (Oneworld Publications, 2009).

12. This is basically the idea behind deism. It uses terms such as Supreme Being, Divine Watchmaker, Great Architect of the Universe, and Nature's God (e.g., in the US Declaration of Independence).

secondary causes and lets those do their work. It is only thanks to the Primary Cause that creatures can become secondary causes. God is not a deity like Jupiter or Zeus—not a being stronger than other beings and superior to all other beings, yet acting like all other beings. Instead, he is the very source of all being—the Absolute Ground of all that happens to exist. This Primary Cause is un-caused, not self-caused, but the Source of all being; not some super-being among other beings, who acts like other beings, but an Absolute Being; not a cause prior to and larger than other causes, but a Primary Cause.

Without a Primary Cause, there could not be any of the secondary causes that science deals with. Why do we need a Primary Cause? The answer is very straightforward: We are contingent beings who could easily not have existed, as the reason for our existence can not be found within ourselves, so we depend for our existence on an overarching, transcending "ground"—a "First Cause," in terms of Thomas Aquinas, or an Absolute Ground, if you will.

To explain this with a different image, think of God as the "ground" and "source" of all that exists in terms of a framework surrounding a spider's web. Likewise, God as the ground of my being is the one who supplies the "framework" that supports the "web of my life" and the "web of this universe." It is in God that we are grounded—otherwise we could not even exist. God provides the "framework" in which "we live and move and have our being," says the apostle Paul.[13] God is the answer to all those questions science has no answer for. Science can only explain one thing in terms of another thing. In doing so, it can only account for incomplete, secondary causes of what exists, but it cannot fill its own gaps; only God offers us the Complete Cause of all that happens to exist. Without God, the universe remains incomplete, as it calls for an Absolute Ground outside itself—the Great Unknown.

Because of this, God does not compete with any creature, not even my parents. My parents are the natural cause of my being here—a secondary cause, that is—but God remains the transcendent Primary Cause of my being in existence and alive. That is why we could say that children come through us but not from us. Even our own children are not "our own," but

---

13. Acts of the Apostles 17:28.

they come ultimately from God, the source and ground of all that exists—otherwise they could not exist, nor could we.

God is "part" of everything in this universe, but without being a physical "part" of it. To put it in a nutshell, God is the First Cause who operates in and through secondary causes. God did not carve out the Grand Canyon single-handedly, but he let the river waters do such a thing. God is the "cause" of all causes, he is even the Cause behind Mill's law stating it is a law that every event depends on some law. Now we know where that law ultimately comes from—it must have come from the Primary Cause, the Great Unknown, the Cause of all causes.

## Big Bang or Creation?

If we have to believe some of the leading physicists and astrophysicists nowadays, then there is not much hope for us in finding a Creator behind this universe. Their reasoning is that the Big Bang story is the modern replacement of the "old" creation story. However, we have also seen earlier that there is good and bad reasoning in human rationality. Arguably the science behind the idea of a Big Bang is pretty good, but the philosophy it is wrapped in is blatantly poor.

How can scientists be so blinded then? Well, scientists acquire their expertise through a rigid scientific education in applying the scientific method and in using sophisticated instruments and very advanced lab techniques in some kind of test-tube setting. That is how they learn their standards, by solving "standard" problems, performing "standard" experiments, and eventually by doing research under a supervisor already skilled in doing this.

Aspiring scientists become gradually acquainted with the models, the methods, the techniques, and the presuppositions of that particular scientific field. Because of their training, scientists are typically unable to articulate the precise nature of the field in which they work until a need arises to become aware of the general laws, metaphysical assumptions, and methodological principles involved in their research. But usually such an awareness remains beyond their scope. That is where we need a bird's-eye view, a meta-view from a meta-level, a science of science if you will—and better late than never.

So let us find out what makes some believe that the creation concept has been replaced by the Big Bang theory. Just listen to what some scientists have to say on this issue. The British cosmologist Stephen Hawking, for instance, talks about the Big Bang in terms of a *spontaneous creation*: "Because there is a law such as gravity, the universe can and will create itself from nothing. Spontaneous creation is the reason there is something rather than nothing."[14] This in turn made the late astrophysicist Carl Sagan exclaim, in the preface of one of Hawking's books, that such a cosmological model has "left nothing for a creator to do."[15] Others, such as the cosmologist Lee Smolin, made sure there is no space for a Creator left by proclaiming that "by definition the universe is all there is, and there can be nothing outside it."[16] Amazing what definitions can do!

I must agree with Smolin that the word "universe" does stand for the whole of all that is material, physical, and visible. So he is right that there can be nothing outside the universe that is physical and visible as well (except for multiple universes perhaps). But this does not mean that there cannot be a non-physical, immaterial, and invisible Absolute Ground that is unconditioned, infinite, and ultimate. As I said before, we are not talking here about a cause prior to and larger than other causes, but a Primary Cause; not a power stronger than and superior to all other powers, but an Infinite Power; not some super-being among other beings, but an Absolute Being. We are not talking here about a "part" of the universe outside the universe, but about the Great Unknown behind all that is, behind all that cannot explain itself.

Next I should point out that the idea of a "spontaneous creation" is sheer philosophical magic; for something to create itself, it would have to exist before it came into existence—which is logically impossible. Imagine, the universe creating itself from nothing! How could the universe "create" itself from nothing—let alone cause itself? It would surely be nice if gold could create itself from nothing, but that is not the way it is in this universe. As the popular saying goes, nothing comes from nothing.

Nevertheless, Hawking tells us that gravity would be able to "create" the universe—spontaneously, so to speak. But I would counter that the

14. Stephen Hawking and Leonard Mlodinow, *The Great Design* (New York: Bantam Books, 2010), 180.

15. Stephen Hawking, *A Brief History of Time* (New York, Bantam Books, 1988), x.

16. Lee Smolin, *Three Roads to Quantum Gravity* (New York: Basic Books, 2001), 17.

law of gravity would have to exist before there can be gravitation. Albert Einstein hit the nail right on the head when he said: "the man of science is a poor philosopher."[17] And there are many more of those. The British physical chemist Peter Atkins, for instance, has the audacity to state that "there is hope for a scientific elucidation of creation from nothing,"[18] after he had said earlier that science has a limitless power and must even be able to account for the "emergence of everything from absolutely nothing."[19] So is there really nothing left for a Creator to do?

Apparently there is quite some confusion as to what the idea of "creating everything out of nothing" means. When the Creator draws everything out of nothingness, he does not push some kind of metaphysical button that starts the Big Bang. Creation does not mean changing a no-thing into a some-thing, or changing something into something else—like chemists change water into hydrogen and oxygen—but it means bringing everything into being and existence. "Nothingness" is not a highly unusual kind of exotic "stuff" that is more difficult to observe or measure than other things; it is not some kind of element that has not found a position yet in the periodic system; it is in no way a material thing that can change into something else, but it is actually the *absence* of anything—and therefore we cannot treat no-thing as a some-thing. Plenty of nothing is still nothing!

Let me clarify this further. Everything that is in "being" makes for an entity; but outside the realm of entities, there is nothing. On the other hand, the problem of "nothingness" is that "nothing" is in itself also an entity, a something. From this we could infer that there *is* nothing, which would imply that nothing somehow *exists*. However, the confusion here is that the notion of nothing is not a real entity but an ideal entity—that is, the conception of missing some real entity (literally, nothing is "being no thing"). Otherwise we would not be able to talk about this nothing.[20] So

---

17. Albert Einstein, "Physics and Reality," *Journal of the Franklin Institute*, Vol. 221, Issue 3 (March 1936).

18. Peter Atkins, *On Being* (Oxford: Oxford University Press, 2011), 12.

19. Peter Atkins, "The Limitless Power of Science," in *Nature's Imagination: The Frontiers of Scientific Vision*, ed. John Cornwell and Freeman Dyson (Oxford: Oxford University Press, 1995), 131.

20. On this complicated issue, see e.g. Joseph M. Bochenski, *The Road to Understanding* (North Andover, MA: Genesis Publishing, 1996), chapter 8.

"nothing" does not really exist, and therefore cannot be the beginning of something else.

When believers of the Judeo-Christian tradition speak of creation "out of nothing" (*ex nihilo*), what do they mean? They certainly are not talking science. Science is about *producing* something from something else, whereas creation is about *creating* something from nothing. Thomas Aquinas made already, centuries ago, the distinction between producing (*facere*) and creating (*creare*).[21] Science is about "producing," about changes in this universe, but creation is certainly not a change; it is not a change from "nothing" to "something." So please note that *creating* something "out of nothing" is not *producing* something out of nothing—the latter of which would be philosophical nonsense. Creation has everything to do with the philosophical and theological question as to why things exist at all, before they can even undergo change. Creation—but not the Big Bang—is the reason why there is something rather than nothing (including something such as the law of gravity).[22]

So we have to make a clear distinction here between "cause" in the sense of a natural change of some kind, on the one hand, and "cause" with a capital C, on the other hand, in the sense of an ultimate bringing into being of something from no antecedent state whatsoever. To be the complete cause of something's existence is not the same as producing a change in something. The Creator does not create something out of nothing by taking "some nothing" and then making something out of it, but he is the complete cause of something's existence. Whereas the Big Bang theory offers us a scientific account of how a later state of the material world might have emerged from an earlier state, creation offers us a meta-physical account of where the material world itself comes from.

Yet, some scientists keep speaking about the Big Bang as if it were the creation of the universe. However, let me say it again, the concepts of "creation" and "Big Bang" are very different from each other: Creation is not some trigger event like the Big Bang. In fact, creation is not an event at all. On the contrary, creation must come "first" before any events, even a Big Bang, can follow. CERN's famous *Large Hadron Collider* may help unravel

21. Aquinas, *De Symbolo Apostolorum*, 33.

22. Aquinas, *De Potentia Dei* 3, 1.

the "mystery" of the beginning of this universe, but certainly not the "mystery" of the creation or origin of this universe. These are two very different kinds of mystery. If the Higgs boson is indeed the particle that could explain why there is mass in the universe, then a physical "mystery" may have been solved—but certainly not the "mystery" of creation, of creating something out of nothing and bringing it into being. When the physicist Leon Lederman called the Higgs boson "The God Particle" in the title of his 1993 book, he was (purposely?) meddling with two kinds of "mysteries."

This discussion has quite some consequences for the way we talk about creation by a Creator. The Creator is the First Cause, uncaused and yet the "cause" or "ground" behind all other, but secondary, causes. Therefore, creation is "first" in origin—we could say "*in* the beginning," like the Book of Genesis does—but not first in time—which would be "*at* the beginning." The Big Bang may be about the *beginning* of the universe, about what happened *at* the beginning, but creation is about the *origin* of the universe, about what is *in* the beginning.

So God, being the Primary Cause, operates in and through secondary causes, including the Big Bang. But the Big Bang does not bring anything into "being"; it works with what is already there in existence. We cannot put these two concepts on the same level, let alone in competition with each other. Creation is very different from what science is searching for. Science can only investigate what already exists. The Book of Genesis, on the other hand, is not a scientific theory of the world's beginning, but rather a monotheistic creed about the world's origin and foundation. It offers a theo-logy of creation, not a sciento-logy.

But there is at least one more important difference. The unfolding of the universe, starting with the Big Bang, is a process that plays in time, but creation cannot follow a timeline, as time itself is a product of creation. Albert Einstein showed us that both time and space are part of the physical world, just as much as matter and energy. Indeed, time can be manipulated in the laboratory. Dramatic time warps occur, for example, when subatomic particles are accelerated to near the speed of light; black holes stretch time by an infinite amount.

Centuries ago, long before Einstein, Thomas Aquinas had already made it very clear that creation is not some chronological episode, located somewhere back in time, when he said, "God brought into being both the creature and time together"[23] and, "Before the world, there was no time"[24] and, "It is idle to look for time before creation, as if time can be found before time."[25] In other words, our universe may have a beginning and a timeline, but creation itself does not have a beginning or a timeline.

So do not place creation *at* the beginning of time, since there is no time until time has been created; creating time "at a certain time" is just tough to do! For us creatures, creation had to begin somewhere, but God himself is time-less. Creation is not something that happened long ago in time, and neither is the Creator someone who did something in the distant past, but the Creator does something at all times—by keeping a contingent world in existence.[26] Creation is something "*in* the beginning"—which made it possible for something to happen "at the beginning."

Whereas the world may have a beginning and a timeline, creation itself does not have a beginning or a timeline; creation actually makes the beginning of the world and its timeline possible. Creation creates chronology, but it is not a part of chronology. Therefore it does not make sense to ask what happened *before* the Big Bang, because there was no time yet until time had been created. Before the Big Bang, there was nothing at all.

Curiously enough, the British cosmologist Stephen Hawking is often quoted as saying that asking what happened before the Big Bang is like asking what lies north of the North Pole. Let me make clear first that Hawking did not say this because he wished to express his belief in creation. Rather he based this on the astrophysical model he developed in collaboration with Jim Hartle, in which the universe has no boundary in space-time. He believes that if we push back the expansion of the universe far enough into the past, space-time is reduced to a single point. Let me explain this.

When you are standing exactly at the North Pole there is indeed no such thing as a direction called "north"; there is no boundary there—it is simply the point where all north-running lines meet and end. In a similar way, there is no such thing as "before," when you get to the point that marks

23. Thomas Aquinas, *Contra Gentiles* II, 32.

24. Thomas Aquinas, *De Potentia Dei* 3, 2.

25. Thomas Aquinas, *De Genesi ad Litteram*, V, 5:12.

26. Thomas Aquinas calls this more specifically preservation or conservation. However, in Aquinas' view, preservation is no different from creation in the sense that they both depend on God (*De Potentia Dei*, 5, 1, 2).

the "pole" of space-time, according to Hawking.[27] Well, I agree with the idea that there is nothing "before the Big Bang," not on physical grounds but on metaphysical grounds: Without creation, there is nothing, not even time. The Big Bang did not "create" time, any more than it "created" gravity, let alone the law of gravity. Creation did!

As a consequence, creation is not a "one-time deal," but it copes with the question as to where this universe ultimately comes from; it does not come from the Big Bang, but may have started with the Big Bang. Without creation, there could not be anything—no Big Bang, no evolution, not even a timeline. Creation sets the "stage" for these and keeps this world in existence. The "rest of the story" would be something for science to tell—and science is definitely trying hard.

What did I myself learn from these reflections? Do not let scientists force us to replace creation with the Big Bang; let them stay in their own territory and make them aware of the philosophical confusion they are causing. The question "*How* did the universe arise?" is very different from the question "*Why* did the universe arise?" The Big Bang answers the how-question regarding creation, whereas creation answers the why-question concerning the Big Bang. They need each other—one cannot exist without the other.

## Evolution or Creation?

In the previous section, I did not try to promote the physical theory of the Big Bang. But if you do accept that theory, as I do, it should never interfere with the notion of God's creation, as long as you use a sound kind of philosophy. In the current section, I would say something similar. I do not want to endorse the biological theory of evolution. However, if you do accept that theory, as I do in some form, it should not interfere with the notion of God's creation, as long as you use a sound kind of philosophy again. So the question is: How can creation and evolution exist together?

---

27. Paul Davies ruminates on the issue of what happened before the Big Bang. His answer is "nothing, not because there exists a mysterious Land of Nothing there but because there is no such place as north of the North Pole." Thus he really misses out on the metaphysical concept of "nothing." ("What Happened before the Big Bang," *God for the 21st Century*, edited by Russell Stannard (Templeton Press, 2000), 10-12).

First of all, we have to stress again that creation is not an event in time. Creation is not about the beginning of this world, but it is about the origin, the ground, and the foundation of this universe, including its beginning and all its subsequent stages. Creation does not start things at the beginning of time but keeps them in existence at all times, in a progression of events. Consequently, the creation account we find in the first chapter of the Book of Genesis is not about what happened *at* the beginning, but about that which everything is based on to begin with—that is, *in* the beginning.

Second, if Thomas Aquinas had known about evolution, he would probably have said that evolution offers us a scientific account of how a later state of the material world might have emerged from an earlier state—whereas creation offers us a metaphysical account of where the material world itself comes from. Creation is about the Primary Cause, whereas evolution would be about secondary causes such as mutation, speciation, and natural selection. Creation creates something out of nothing, whereas evolution produces something out of something else—by following biological laws in the same way as planetary motions follow physical laws. God does not make things himself but he makes sure they are made through his laws; God is the First Cause who operates in and through secondary causes, including the causes of evolution.

How can evolution work within the setting of creation? Thanks to creation, the universe has been outfitted with a "cosmic design" of laws and boundary conditions. Let me briefly explain this idea. What the universe and man-made machines have in common is the fact that they follow the same rule: They are only useful because the laws of nature have been harnessed within certain boundary conditions or restraints. All scientific laws operate within so-called boundary conditions or restraints. There are half a dozen physical constants that are given, not derivable from any known theory (the speed of light, the force of gravity, etc.). All other physical factors are causally determined by these more fundamental constants.

It is not impossible, though, that some of these fundamental constants may in time be derived themselves. At one point, for instance, the boiling point of water was taken as a physical constant, but is now considered the result of quantum mechanical laws. All scientific laws operate within boundary conditions or restraints like these. The law of gravitation, for example, is constrained within the boundaries set by general relativity for a specific average density of matter—otherwise a cosmological catastrophe would be generated. All other physical factors are causally determined by these more fundamental constants.

Had the boundary conditions of this universe been even slightly different from what they actually are, our universe would probably not exist, and neither would we. Had neutrons been just a bit lighter than protons, instead of the other way around, atoms could not exist, because all the protons in the universe would have decayed into neutrons shortly after the Big Bang. Even the atheist and cosmologist Fred Hoyle had to acknowledge this fact: "A commonsense interpretation of the facts suggests that a super-intellect has monkeyed with physics, as well as with chemistry and biology, and that there are no blind forces worth speaking about in nature."[28] Indeed, the design of the universe looks suspiciously like a fix. As the American physicist John Wheeler put it, "A life-giving factor lies at the centre of the whole machinery and design of the world."[29]

Well, this is what I call the cosmic design. It is this *cosmic* design that also explains which *biological* designs are possible by regulating which biological designs are successful in reproduction and survival—as much so as the cosmic design regulates which bridges are successful technological designs. Due to the cosmic design, there is enormous potentiality in nature. All biological designs have to go through the "filter" of natural selection. Only what is in accordance with the cosmic design can be filtered as a successful biological design—successful in reproduction and survival. Natural selection can only select those specific biological designs that follow the rules of the cosmic design (again, designers, engineers, and architects must do the same thing).

What then is this "something" that makes them work successfully and effectively? The answer is that they must have followed the "rules of the cosmic design," which restrict the range of possible end results. Birds must follow the same aerodynamic laws as planes—otherwise they surely fall from the sky—and fish must follow the same hydrodynamic laws as submarines do. To put it in a nutshell, organisms do not have successful

---

28. Fred Hoyle, "The Universe: Some Past and Present Reflections," *Engineering and Science* (Nov. 1981): 12.
29. John A. Wheeler, "Foreword," in *The Anthropic Cosmological Principle*, John D. Barrow and Frank J. Tipler. (Oxford, U. K.: Clarendon Press, 1986), vii.

designs because they have survived; on the contrary, they have survived because their biological designs square well with the cosmic design. They must have "something" in their biological design that carried them through the filter of natural selection.

Leon Kass, a University of Chicago professor and physician, could not have worded it better: Organisms "are not teleological because they have survived; on the contrary, they have survived (in part) because they are teleological."[30] Charles Darwin had it backwards or upside down. He thought he could explain nature's functionality with the process of natural selection, but instead natural selection selects what is functional within the setting of the cosmic design, so it can only be explained by assuming a cosmic design. Functionality is not an outcome of natural selection but a condition for natural selection. Natural selection does not explain functionality but uses functionality in order to select.

This may clarify why evolution is not as haphazard and undirected as some biologists think it is. Indeed, there is randomness in evolution: Mutations are supposed to be random, environmental changes are considered to be random, and natural selection allegedly acts randomly. However, the term "random" can carry many different meanings, as it does in the previous examples. Each time a biologist says that something happened "randomly," we should ask: How do you mean? When they consider mutations random, they have a different meaning in mind than when they call natural selection random, because the latter process is actually highly selective, not random; and yet it is random in the sense of opportunistic and short-sighted with regards to *future* needs, but definitely not random in the sense of arbitrary, given the fact that natural selection has a definite "preference" for what fits best—"the fittest."

In case of mutations, however, the term "random" has a very different connotation. To begin with, mutations in genetic material are considered random in the sense that they are (as far as we know) "unpredictable" as to when and where they strike. We may know more nowadays about what causes them (factors such as radiation, chemicals, free radicals, etc.), but we still cannot predict at what location in the DNA those factors will hit and what changes they might generate there. So they may not be principally but

30. Leon R. Kass, "Teleology, Darwinism and the Place of Man: Beyond Chance and Necessity?" *Toward a More Natural Science* (New York: Free Press 1988), chapter 10.

practically unpredictable. Second, mutations are random in the sense of "arbitrary," because mutations do not select their target but hit in-discriminatorily. Third, mutations are also random in the sense of "opportunistic," because they occur regardless of any future needs. There is no physical mechanism that detects which mutations would be beneficial and then causes those mutations to occur. But that is where randomness ends. In its most general sense, it is a statistical concept that can be scientifically tested for; all it tells us is that causes can be independent of each other.

What all the above distinctions have in common is a scientific, statistical connotation. Random occurrences are not predictable, yet they exhibit regularity in the aggregate after many repetitions, so there is a certain probability or likelihood of the events' occurring. This means that the concept of randomness has a narrow technical meaning having to do with the statistical correlations among things and allowing us to calculate probabilities. Events do not happen "by chance" because they are uncaused or we cannot trace their causes, but because there are so many causal chains that intersect with each other. From this follows one very important caveat: Never confuse randomness with fate or doom, as these latter two terms are not scientific concepts but world-view notions.

However, even though there is some randomness in evolution, the evolutionary process as such is not completely haphazard or undirected. Let me use an analogy to explain this a bit further. A river follows a "path of least resistance" according to the "topographic design" of the landscape. In a similar way, the "stream" of evolution follows a path somehow regulated by the cosmic design of our universe. In other words, evolution follows the "path of least resistance" in the landscape of the cosmic design. It does not just flow in a random way. Further evidence can be found in the fact that marsupials in Australia and tenrecs in Madagascar have evolved into groups that resemble animals on the mainland as diverse as hedgehogs, shrews, opossums, mice, porcupines, and even otters—although they are not closely related to any of these groups, which indicates they must have gone through their own, parallel evolution. Although there was some randomness involved, both groups evolved in a very similar way.

The cosmic design also explains how there can be so much potentiality in this universe. The Belgian biochemist, cell-biologist, and Nobel Laureate Christian de Duve describes the origin of life as follows: "The pathway followed by the biogenic process up to the ancestral cell was almost entirely preordained by the intrinsic properties of the materials involved, given a certain kind of environment or succession of environmental conditions."[31] Evolution seems to have a time arrow moving from "less complex" to "more complex," but make no mistake: The simple is in no way less orderly than the complex; even a simple snowflake actually shows a very intricate order.

In other words, there is no reason to think that evolution builds order out of pre-order or even chaos. The order we see in nature does not and cannot come from chaos, but must come from a more fundamental, pre-existing order at a deeper level—which is the cosmic order and design of creation, harnessing everything that takes place in the universe, including the process of evolution. Compare this with the situation where shaking a jar of variously shaped candies would not create much more order, whereas shaking a jar with *round* candies definitely would, because those candies have an underlying order (at a "deeper" level") that allows for a "hexagonal closest packing" structure.[32]

In a similar way, the cosmic order regulates which processes are physically possible and which designs are biologically fit. The Nobel Laureate and microbiologist Werner Arber put it this way: "How such already quite complex structures may have come together, remains a mystery to me. The possibility of the existence of a Creator, of God, represents to me a satisfactory solution to this problem."[33] I would interpret this in terms of a Creator God behind the cosmic design, which creates the "bed" in which the stream of evolution meanders.

Thanks to the cosmic design, the universe begins to look more like a great "thought" of the Creator than some kind of a mindless machine. That is what we learned earlier regarding the physical world, but it also applies to the biological world. The cosmic design regulates which biological designs are successful in survival. Take, for instance, the foraging

---

31. Christian de Duve, *Blueprint for a Cell: The Nature and Origin of Life* (Burlington, NC: Neil Patterson Publishers, 1991), 214.

32. This example is used by the University of Delaware physicist Stephen M. Barr, "Fearful Symmetries," *First Things*, Oct. (2010).

33. Werner Arber as quoted in Henry Margenau and Ray Abraham Varghese, eds., *Cosmos, Bios, Theos: Scientists Reflect on Science, God and the Origin of the Universe, Life and Homo Sapiens* (La Salle, IL: Open Court, 1992), 142.

behavior of animals:[34] Natural selection promotes foraging patterns that maximize or optimize the net calorie intake. As a result, organisms forage first for food items that give the greater harvest per unit time, so they get maximum output for minimum input.

Does this not look like there is logic and rationality in nature? I am not saying that there is logic and rationality in the brains of these animals, since they do not make conscious decisions. These organisms do not calculate costs and benefits, but they act *as if* they make strategic decisions. It is natural selection that "promotes" the kind of behavior that squares with the cosmic design. And what then is this cosmic design based on? It has its origin in the "logic and rationality" of its Maker. And I would add that this would also explain the logic and rationality of human beings.

As it turns out, there need not be any conflict, contradiction, or incompatibility between creation and evolution; they actually need each other and complement each other—that is to say, as long as we keep our philosophy straight. I am not speaking here in terms of "reconciliation," as if science and religion were just two different ways of expressing single truth. They in fact convey two very different kinds of truth. Attempts to "reconcile" the data of religion with the data of science usually go only one way—"interpreting" the data of religion so as to leave the data of science intact (as if science is always right). Many believe this is what ultimately happened after the "Galileo affair." They consider the case exclusively in terms of either-or. If Galileo was right, the Bible and the Church must have been wrong. Or, if the Bible and the Church were right, Galileo must have been wrong.

In reality, though, the situation was much more complex. On the one hand, we found out that Galileo's "facts" were not as clear as he portrayed them. People like Cardinal Bellarmine actually called for more evidence to support the case.[35] In addition, the frequency and acidity of Galileo's attacks against other astronomers played an important role in causing many Jesuit astronomers of the Roman College to withdraw their initial support

34. Alan C. Kamil, John R. Krebs, and H. Ronald Pulliam, *Foraging Behavior*, (New York: Plenum Press, 1987).

35. Cardinal Bellarmine wrote in a letter to Fr. Foscarini, "If there were a true demonstration that ... the sun did not go around the earth but the earth went around the sun, then it would be necessary to use careful consideration in explaining the Scriptures that seemed contrary."

of Galileo—which he later would need so badly. On the other hand, the Church's position should have been more clear-headed. The Church could have learned from Augustine that the Book of Scripture is not the Book of Nature, and is not about science but about salvation. Even Aquinas' teacher, Albert the Great, had always promoted a peaceful coexistence of science and religion—each having its own authority.[36]

But no matter what the actual situation was, both parties in the "conflict" did learn from this painful experience that science reads the Book of Nature, whereas the Church reads the Book of Scripture. Those two books complement each other, for they have the same Author—a match made in Heaven. To set them up against each other is a false dichotomy. Cardinal Baronius worded it well at the time, "The Bible teaches us how to go to heaven, not how the heavens go."[37] In his speech of Oct. 31, 1992, John Paul II rightly spoke of a "tragic mutual incomprehension."

However, this historical lesson can easily be forgotten. At the very moment scientists decide to package evolution in a poor philosophical wrapper, they may *force* a conflict again between science and religion, thus making us believe that if science is right, religion must be wrong. In response I would counter: Why would there be a conflict between these two? As Albert Einstein, a scientific authority, once said, "Science without religion is lame, religion without science is blind."[38] Or as Pope John Paul II, a religious authority, put it, "Science can purify religion from error and superstition. Religion can purify science from idolatry and false absolutes."[39]

Religion must keep itself open to the findings of science, and science must respect what lies outside the boundaries surrounding its inquiries.

36. Albert the Great said "The aim of natural science is not simply to accept the statements of others, but to investigate the causes that are at work in nature," *De Mineralibus*, II, II, I.

37. This remark, which Baronius probably made in conversation with Galileo, was cited by the latter in his *Letter to the Grand Duchess Christina* (1615).

38. Albert Einstein, *Science, Philosophy and Religion, A Symposium* (New York: The Conference on Science, Philosophy and Religion in Their Relation to the Democratic Way of Life, 1941).

39. Address to the *Pontifical Academy of Science*, October 31, 2008. And in a letter to George V. Coyne, SJ, Director then of the Vatican Observatory, June 1, 1988, in Robert J. Russell, *John Paul II on Science and Religion: Reflections on the New View from Rome* (Rome: Vatican Observatory Publications, 1990).

Together they make us whole, whereas on their own, they are incomplete and underperforming. The question "*How* did life arise?" is very different from the question "*Why* did life arise?" Evolution answers the how-question regarding creation, whereas creation answers the why-question regarding evolution. Both questions are legitimate and call for an answer. But be assured, "we're never going to find anything that's outside of God's realm," according to Jennifer Wiseman, Chief of the ExoPlanets and Stellar Astrophysics Laboratory of NASA Goddard Space Flight Center.

Obviously, we are dealing here with two disparate domains that should be separated in a way analogous to the separation of church and state. I strongly advocate a "separation" of science and religion, with each one having its own authority and expertise. But make no mistake, this is a separation of two institutions, not two dimensions of life. Science should stay away from religious territory in attempting religious interventions—as much as religion should stay clear of science in attempting scientific interventions. Nevertheless, science should protect religion from errors, and religion should keep science within rational and moral bounds. Render to science what is material, measurable, and visible, but do not render to science what is God's. So honor the "fences," but do not forget that, no matter on which side of the fence you happen to be, you do have a "neighbor."

Confusion sets in when people break the fence and replace the theory of evolution with the doctrine of evolutionism. Like most other "isms," evolutionism is a one-sided, monopolistic ideology that exclusively focuses on one very specific perspective on life by proclaiming that this outlook makes you see everything—that is, literally, all there is. Most "isms" love to simplify the vast complexity of reality by substituting reality with one of its simplified models or maps. They suffer from some form of megalomania in their demand for absolute authority. They have an autocratic belief in the omni-competence of their scientific field and their own narrow scientific approach. That is where science becomes a semi-religion.

When we inflate neo-Darwinism and turn it into evolutionism, it becomes a real "ism" that claims to have an explanation for everything in life. Proponents of this ideology glorify the theory of evolution and use it as the exclusive foundation of their worldview: Natural selection is all there is and all that counts, so they believe. Therefore, they want to silence religion, because they reject creation and worship evolution instead: Everything in nature is the result of natural selection, so they say—yes, everything, including religion.

But when they make such a claim and down-grade religion to the status of being also the mere product of natural selection, they forget that the very theory of natural selection must then also be the mere product of natural selection. I would say such a doctrine ultimately devours itself. Imagine, natural selection producing the theory of natural selection! That would be like the miracle of a hand drawing itself; a hand may draw a picture of a hand, of course, but a hand that draws a hand cannot produce the very hand that does the drawing! The very hand that does the drawing must be "more" than the hand that is being drawn on paper. Likewise, the mind that came up with the theory of natural selection must be more than a product of natural selection.

In contrast, when the famous geneticist Theodosius Dobzhansky, a dedicated member of the Orthodox Church, said that nothing makes sense in biology except in the light of evolution,[40] he was very well aware of the fact that there is more to life than biology. If neo-Darwinism claims that most biological facts do not make sense without assuming evolution, religious faith has equal rights to claim that evolution does not make sense without assuming creation. We may have come here through evolution, yet we came here from God. Let us face it, the theory of evolution does not answer all our questions, and it certainly does not answer above all the great philosophical question: From where does everything come?[41]

Let us conclude this chapter with one sentence: If everything ultimately comes from creation, we may not be alone; there may be someone behind and beyond this universe, someone beyond and behind all that we see through telescopes and microscopes. If so, our universe may not be a God-forsaken place after all. But the question still remains: Do we really *need* any help from "beyond"—not to mention the question as to whether we can *expect* any help?

---

40. Theodosius Dobzhansky, "Biology, Molecular and Organismic," *American Zoologist*, vol. 4 (1964): 443–452.

41. For a more detailed discussion see Gerard M. Verschuuren, *God and Evolution? Science Meets Faith* (Boston, MA: Pauline Books, 2012).

# 4. Cues from Beyond

Our universe seems to be incomplete in itself and calls for someone behind and beyond our universe—someone somewhere beyond the black holes. Even if we believe this to be true, we could still ask ourselves whether we would really *need* any help from there. Those who say "God, I do not need your help," usually need God's help more than ever, I found out. God is at the basis of all our deepest questions—questions such as "What is our origin?"; "What is our purpose?"; "Where does everything that exists come from and where will it end?"

I am going to simplify and unify the above questions into one single, more philosophical question: Why is there something rather than nothing? In order to explain something that could have not existed, we need an Absolute Being that necessarily exists and is perfect in itself, so we found out.

In more technical terms: Only an Absolute Being can explain the existence of contingent beings. The existence of the contingent depends completely on the existence of the Absolute. The universe is contingent, and is therefore incomplete without the existence of the Absolute. Well, this Absolute is what most believers would call God, their Creator, or the Great Unknown.

## Is There Anyone out There?

There certainly is someone behind and beyond our universe—as I have been discovering more and more clearly myself. So what has made me so sure in the end?

Let me immediately make clear that my certainty is not based on so-called *proofs* of God's existence. Please do not take this statement the wrong way; I am not saying they are useless or even impossible, but they are not as error-proof as many think. Why? Well, they are not really proofs in a logical, mathematical, let alone scientific sense. So what are they then? Thomas Aquinas mentions five of them, but it is very telling that he does not call them "Five Proofs" but "Five Ways."[1] What do they state and what do they achieve?

Let me begin with the question as to why our universe is the way it is, or why it even exists at all. The cosmologist Stephen Hawking worded the question this way: "You still have the question: why does the universe bother to exist?"[2] That is a good philosophical question from a non-philosopher. As a matter of fact, our universe need not be the way it is, and it need not even exist, so we found out. In other words, our universe is neither necessary nor absolute, but contingent instead—that is, finite and dependent.[3] However, if there is no inherent necessity for the universe to exist, then the universe is not self-explaining and therefore must find an explanation outside itself. Obviously, it cannot be grounded in something else that is also finite and not self-explaining (that would lead to infinite regress[4]), so it can only derive from an unconditioned, infinite and ultimate ground, which is a Creator God—a Being not self-caused but un-caused.

This is actually the "Third Way from Contingency" that Aquinas mentions: "Contingent" means that they do not have to exist; but since they do exist, there must be a necessary Being that causes them to exist. Well, this Being is what we call God, according to Aquinas. The four other Ways (or "arguments" of God's existence) that Aquinas mentions are essentially variations of this one particular Way. But let me underscore that even Aquinas did not have the presumption of logically proving God's existence; his reasoning is not concluded by the sentence, "Therefore, God exists or must exist (QED)" but by the less overreaching statement, "And this all think of

1. The *Quinque Viae* from the *Summa Theologiae* are as follows: 1. The Argument from Motion, 2. The Argument from Efficient Cause, 3. The Argument from Contingency, 4. The Argument from Gradation, and 5. The Argument from Design.
2. In interviews with Gregory Benford, "Leaping the Abyss: Stephen Hawking on Black Holes, Unified Field Theory and Marilyn Monroe," *Reason*, 4.02 (April 2002): 29.
3. It is only in the dialectical materialism of Marxism that matter is considered to be the absolute.
4. A series of propositions leads to *infinite regress* if the truth of the first one requires the support of a second one, the truth of the second one requires the support of a third one, and so on, and the truth of the next-to-last one requires the support of the last one, where the number of "last" approaches infinity.

as God."[5] Aquinas is more or less clarifying what we mean when we speak of God, and what this entails. This should not surprise us since, for religious believers, faith is not a conclusion to a logical syllogism, but is ultimately a divine gift from beyond. Besides, in the society Thomas lived in, there were no real self-proclaimed atheists, so he did not have to prove anything for skeptics. For long, it was just unthinkable to deny God's existence until it became a viable option in more modern times, when new ways of understanding the world were promoted as new ways of dealing with God.

Although the Five Ways of Aquinas are not presented as conclusive proofs of God's existence in a logical or mathematical sense, they still should be taken in a *rational* sense: They offer strong reasons or arguments for God's existence, which makes them work like powerful pointers to a Creator God as the best possible (and probably only) rational explanation for the fact that this universe does exist and is the way it is.[6] The fact that there is any world at all, or any causality at all, or any law at all, is the proper starting point for an argument for God as Primary Cause.

All Thomas' arguments are philosophical clarifications of what we mean when we speak of God and what the consequences are of the way we understand this notion of God. It tells us what we do mean and what we do not mean when we speak of God. We either accept that nothing explains itself and leave it at that—which is basically irrational—or we restore rationality by stating that nothing explains itself and therefore needs an explanation "beyond" itself.

I would say the latter choice is the more rational one of the two, assuming we want to walk a road to understanding—otherwise we would have to conclude that this universe is absurd. Again, this rational approach is not a proof with mathematical or logical certainty, since it is based on the assumption that the universe can be rationally understood, but it surely is a very powerful rational argument in favor of God's existence. It actually qualifies what we mean and do not mean when we speak of God. Some people, on the other hand, think that when we begin to use reason, we have no choice but to abandon faith; conversely, they think that if we have faith, we cannot use reason. In fact, faith and reason can, must, and do live

5. Aquinas, *Summa Theologica* I, 2, 3.

6. The late Harvard philosopher and mathematician Alfred North Whitehead calls this the "principle of concreteness."

in harmony. We should be faithful in our reasoning and reasonable in our faith—even when, or specifically when, it comes to God. We cannot live by faith alone neither by reason alone, but by a harmonious combination of faith *and* reason.

Besides, I would like to mention someone who came into this discussion from an entirely different angle. The famous mathematician Kurt Gödel from Princeton University rigorously and mathematically proved in his so-called incompleteness theorem[7] that coherent systems are incomplete (while complete systems are incoherent). As to our discussion, no coherent system—not even the system of science—can be completely closed; any coherent system is essentially incomplete and needs additional "help" from outside the system, according to Gödel. His theorem proves that science can never fill its own gaps. To "prove" the consistency of the system requires an "act of faith."

Gödel even went as far as believing that a credible account of reality itself cannot be completely closed and would therefore be incomplete if we do not invoke God.[8] Gödel was said to be very cautious in mentioning this belief in scientific circles, because he considered it potential dynamite. But what he did tell us is that our capacity to know truth transcends mere formal logic; in other words, there are truths that we cannot "prove" in a logical way.

There is a more serious problem, though: How can the Great Unknown ever be known? Is that not a *contradictio in terminis*? Could an infinite, transcendent God ever be accessible to the finite human mind? It depends. Some people, such as Immanuel Kant, declare God unknowable ahead of time. Kant attempted to merge the ideas of rationalism and empiricism. Rationalism holds that there is certain innate knowledge within everyone. On the contrary, empiricism maintains that we are born as blank slates, and all knowledge is gained by experience.

Kant pulled them together and concluded that the content of knowledge comes by experience (empiricism), but the structure or form of knowledge is developed in the mind (rationalism). Hence, if one can only know something through experience by means of the senses, and if that sensed knowledge can

---

7. Francesco Berto, *There's Something about Gödel: The Complete Guide to the Incompleteness Theorem.* (Hoboken, NJ: John Wiley and Sons, 2010).

8. See David Goldman, "The God of the Mathematicians," *First Things*, 205 (2010): 45-50.

only be structured in our minds by innate categories, then we can only know things as they *appear* to us; we can never know reality as it actually is.

My first response to Kant would be that he seems to know at least one thing about the Unknown—that it is unknowable! But do we not actually know more than just one thing? Can we not make the Great Unknown a bit more known? If we could never say anything right about God, we could never say anything wrong either. Why would it be flawed to think that the Great Unknown can be known (to a certain extent) based on what we see around us combined with the power of reason—unless we give up on reason ahead of time?

Remember Plato's paradox: "How would you search for what is unknown to you?" We are in search of something "unknown"—otherwise we wouldn't need to search anymore—and yet it must be "known" at the same time—otherwise we wouldn't know what to search for, or would not even know if we had found what we were searching for. One could even make the case that if our minds are a reflection of God's mind, created in his image, then our minds can be a gate towards God, making God somewhat accessible to the human intellect through the natural light of reason, which is a gift from God.

The problem with Kant is that he heavily relied on the Scottish philosopher David Hume. Hume was one of the first skeptic philosophers, who questioned the very idea of objective truth. He argued that all connections we observe in life are nothing but constant conjunctions in our minds, and our perceptions of them never give us insight into the *modus operandi* of the connection. So he declared causal connections to be mere "metaphysical" inventions, based on an illusion. As a consequence, causal connections in themselves are ultimately subjective phenomena on Hume's view.

What is wrong with his analysis? Although we do know the world through sensations or sense impressions, they are just the media that give us access to reality. There is a way things are, independent of how they may be apprehended. The philosopher John Haldane put it well when he said, "One only knows about cats and dogs through sensations, but they are not themselves sensations, any more than the players in a televised football game are color patterns on a flat screen."[9] Knowledge does rest on sensa-

9. John Haldane, "Hume's Destructive Genius," *First Things*, 218 (2011): 23-25.

tion, but this does not mean it is confined to it.

If we were to follow Hume's philosophy, we would end up with what the late physicist and historian of science Stanley Jaki calls "bricks without mortar." He says about Hume's sensations, "The bricks he used for construction were sensory impressions. Merely stacking bricks together never produces an edifice, let alone an edifice that is supposed to be the reasoned edifice of knowledge."[10] In short, Hume leaves us in a cognitive desert, and so does Kant.

This takes us back to Aquinas: "We must believe that God exists, which is clear by reason."[11] What Aquinas means by this statement is that *reason* leads us to assume a Primary Cause, whereas *faith* in God discovers that this concept refers to the God of our faith. This does not mean that reason can fully fathom God, for elsewhere he says, "We cannot understand what God *is*, but what he *not* is."[12] In other words, our (philosophical) knowledge is limited to knowing *that* God is, more so than *what* he is. Yet, reason tells us that this universe cannot explain itself, but needs an ultimate non-contingent explanation. A relative explanation for a relative world would only shift the problem; it is only the Absolute Unknown that can explain the relative, the known. In order for the Great Unknown to be the Absolute, it must be unchanging and infinite—all-knowing, all-powerful, and all-present.

I would say the Great Unknown is beginning to take on some "shape." In addition I would say that, if it is true we were made in God's image, then we can know more about God when we know more about ourselves.

It should not surprise us that some people are not convinced by the rational arguments I brought up regarding the existence of a Creator God. If they want to give up on rationality and the power of reason, I cannot fight them—for one cannot use rationality to fight irrationality. If they want to stay in limbo, agnosticism is their last resort, for it asserts that we just do not know whether God does exist or not, and what is worse, we have no way of ever knowing one way or the other. Agnosticism sounds harmless

10. Stanley Jaki, *The Road of Science and the Ways to God* (The University of Chicago Press, 1978), 103.

11. Aquinas, *Compendium Theologiae*, 3.

12. Aquinas, *Summa contra Gentiles*, I, 30.

and safe by claiming that logic cannot demonstrate the *falsity* of a belief in God—it is said to be an unbeatable, unverifiable hypothesis—but neither can it demonstrate the *truth* of a belief in God—it is said to be a daring, undecided hypothesis.

But I wonder if agnosticism is really that harmless. Very often agnostics do have a rather negative and highly selective attitude towards religion and religious people in particular—they just declare them "stupid." Agnostics often think that their own logic is so compelling that everyone who disagrees with their agnostic conclusions must be misinformed or just brainless. Is it really "not a creed but a method," as Thomas H. Huxley, who coined the term 'agnostic,'[13] once pretended? To be sure, as a method, it boils down to a "philosophy of the unknowable," but very often it excludes only *religious* truth from the domain of knowledge; it limits the power of reason to the domain of scientific truths only, at the exclusion of religious or philosophical truths. So it would be scientism in disguise.

Interestingly enough, this is a contradictory claim, as agnosticism in itself is not a scientific truth either. Besides, this demand for scientific evidence through laboratory testing is in effect asking God, the Primary Cause and Supreme Being, to become our servant. As Benedict XVI put it, "The arrogance that would make God an object and impose our laboratory conditions upon him is incapable of finding him. For it already implies that we deny God by placing ourselves above him."[14] One could argue that the question of God should be treated differently from other knowable objects in that this question regards not that which is "below" us, but that which is "above" us. We cannot submit God to experiment by testing him just as theories and products are tested. We cannot submit him to the conditions we deem necessary if we are to reach certainty. Benedict goes on to say, "The highest truths cannot be forced into the type of empirical evidence that only applies to material reality."[15] Therefore, I think Lenin was quite on target, for a change, when he said that "agnosticism serves as a fig-leaf for materialism."[16]

---

13. T. H. Huxley, "Agnosticism: A Rejoinder." In *Collected Essays*, vol 5, *Science and Christian tradition.* (London: Macmillan, 1889).

14. Joseph Ratzinger, *Jesus of Nazareth* (San Francisco, CA: Ignatius Press, 2007), 36.

15. Ibid., 216.

16. Vladimir Lenin, "Materialism and Empirio-Criticism," in: *Lenin, Collected Works,*

Could we ever prove then that there is *no* God? G.K. Chesterton once said, "Atheism is the most daring of all dogmas, for it is the assertion of a universal negative."[17] Chesterton is right; it is much easier to establish that there is a black swan somewhere on the earth than to prove that there is not one at all. We may perhaps validly conclude that God is unknown (as agnosticism asserts), but it is very hard, if not logically impossible, to conclude that God is in fact absent (as atheism claims). Atheism often amounts to some form of circular reasoning: "There is no God; so all proofs for the existence of God must be flawed because there is no God."

It is just impossible to close a search for God with the conclusion that there is *no* God. No searches ever conclusively reveal the *absence* of their object. Likewise, it is impossible to prove that some accident has no cause at all; the real cause may still be eluding us—and so may God. Let us not forget that absence of evidence is not evidence of absence.[18]

Whereas agnosticism regards God-talk unverifiable but nonetheless meaningful, some people maintain that, if unverifiable, God-talk is *ipso facto* meaningless. In their view, language is supposed to be either empirical (dealing with observable facts, like in science) or logical (adhering to rules, like in logic and mathematics), otherwise it is considered a nonsensical abuse of language. The fundamental problem with approaches like these is that they determine ahead of time the outcome they desire to see. They define what is legitimate by making sure they exclude what they do not wish to be legitimate. And besides, you wonder whether this claim in itself is either logical or empirical—or merely meaningless. Some people just have this contemptuous habit of dismissing as meaningless those concepts whose meanings elude them.

And then there are those atheists who declare religion, even if there is anything "empirical" about it, a mere product of the mind, a form of wishful thinking, so they say. They like to take God's seat on the very throne they had declared vacant. Hence, religion ends up being an illusion or delusion. When religion tells us that we were made in God's image, these atheists would point out that it is actually reversed: God was made in our

---

(Moscow: Progress Publishers, 1972), Volume 14, 17-362.

17. G.K. Chesterton, *Twelve Types* (Middlesex, U.K.: The Echo Library, 2008), 29.

18. I borrowed this line from the British cosmologist and astrophysicist Martin Rees.

own image. Sigmund Freud, for instance, was of the opinion that the adoption of religion is a reversion to childish patterns of thought in response to feelings of helplessness and guilt. We feel a need for security and forgiveness, and so invent a source of security and forgiveness—God. In this view, religion is seen as a childish delusion, whereas atheism is taken as a form of grown-up realism.

Did atheists such as Freud effectively refute religion? First of all, if Freud claims that basic beliefs are the rationalization of our deepest wishes, would this not also entail that his own atheistic beliefs could be the rationalization of his own desires as well? Don't we often think what we wish? Don't some people, like ostriches, choose to deny what they fear or hate? Second, even if belief in God were wishful thinking, one could never prove that it is nothing more than wishful thinking. When Freud says there is no God, Freud seems to be his prophet! The God one would like to exist may actually exist, even if the fact one wishes it encourages suspicion. Third, the claim that the brain created God could easily be countered with the reversed claim that it was God who created the brain. If so, we do not project a human image onto heaven, but heaven shows us what we as human beings can and should be like. No wonder that Albert Einstein said about atheists that they are "creatures who—in their grudge against the traditional 'opium of the people'—cannot bear the music of the spheres."[19]

Then there are also atheists who are just hard-core skeptics à la Hume, who deny the validity of nearly all aspects of knowledge, because we are not supposed to know any truth with certainty. In contrast, I would say scientists do need a critical mind, but not a skeptical mind. Skeptics find a flaw in every truth claimed. Skepticism makes for a very restrained view on the world—actually so restrained that absolute skeptics cannot even know whether they have a mind to doubt with. And yet, skepticism rejects any authority while declaring itself as the new authority. Let us declare that God is dead, then we ourselves will be God.

In response I would say that skeptics turn things the wrong way. We often do need to eliminate errors to get to the truth; yet, our ultimate goal is not to avoid errors but to gain truth. We want to know—not to know what we do *not* know. Skeptics, on the other hand, make it their final goal to avoid

19. Quoted in: Max Jammer, *Einstein and Religion: Physics and Theology*. (Princeton: Princeton University Press, 2002), 97.

errors, in denial of the fact that eliminating errors is only a means to gaining truth—so they end up with an empty shell of complete mistrust. In general, I would say that the objections atheists have *against* God's existence are at least as weak as, if not much weaker than, the rational arguments *for* God's existence. I am afraid they give atheism only a bad name (take this as a joke, please)! I hope you agree that we should not spend more time on them.

Now that we have found a way to talk about a Creator God, we have also found a new "entry" into many other issues discussed in this book earlier. Let us start with the fact that there is a "cosmic design" behind this universe. How come this world is so orderly, instead of being completely chaotic?

Let us face it: If there were no *order* in the universe, it would not make sense to search for laws of nature in physics, chemistry, biology, and other disciplines. It is due to this orderly design that we can explain and predict—which would not be possible in a world of disorder and irregularity. Since we may assume that like causes produce like effects, we are able to explain and predict. So the question is: What then explains that the universe is so orderly, and is based on laws of nature and certain boundary conditions? Where do all our scientific laws come from? Why is this universe law-abiding?

The answer to this question is that the cosmic design is contingent and therefore needs an explanation beyond itself—unless we want to give up on rationality ahead of time. The only rational explanation would be the existence of an Intelligent Designer. In this line of thought, the order behind this universe is basically a given, somehow imprinted on the universe at its birth, and fixed thereafter.

Science can never prove there is order in this universe, but must *assume* it. The tool of falsification,[20] for instance, is in fact based on this very assumption as well: The fact that scientific evidence can refute a scientific hypothesis is only possible if there is order in this universe. Without order, there would not be any falsifying evidence. When we do find falsifying evidence, we do not take that as proof that the universe is *not* orderly,

---

20. Karl Popper is the famous defender of falsifiability and testability in science: "Every 'good' scientific theory [...] forbids certain things to happen." (*Conjectures and Refutations: The Growth of Scientific Knowledge* (London: Routledge and Keagan Paul, 1963, 33-39).

but as an indication that there is something wrong with the specific order we had conjectured up in our minds.[21] Falsification is essentially based on order and cannot be falsified by disorder, so counter-evidence does allow us to falsify theories, but not the principle of falsification itself. In utter amazement, Albert Einstein wrote in one of his letters, "But surely, *a priori,* one should expect the world to be chaotic, not to be grasped by thought in any way."[22] Einstein was enough of a philosopher to realize the importance of a given order, one of the main pillars of science.

When I hear John Stuart Mill say, "It is a law that every event depends on some law," I always wonder whether he ever wondered where that law then came from (other than from himself). Indeed, it is a fundamental "rule" in science that everything it tries to explain or predict is based on laws—laws of physics, biology, chemistry, and so on. But the fundamental question is where such a "rule" comes from. The "law" that everything is based on laws definitely does not come from science itself, for science cannot explain its own rules, any more than chess can explain its own rules. Science can explain things by using laws, but it certainly cannot explain the very existence of those laws. So where do those laws come from then? I do not see any other answer than this: The "law and order" of our universe is not an *a priori*—rooted in the way we think about the world, à la Kant—but it is a given, rooted in the way the world *is* due to creation. Without God, there would only be mere chaos, or actually nothing.

Something similar holds for the notions of design and functionality, especially so in the life sciences and the technical sciences. A heart and a pump, or an eye and a camera, can and do work because of the way our universe has been designed and outfitted with teleology. They are "successful" designs, given the way the cosmic design is—and that makes them functional. Without some kind of cosmic design in the background, they could not work at all. As a consequence, natural selection can only select those biological designs that are in accordance with the cosmic design.

---

21. Albert Einstein said something to this effect: "No amount of experimentation can ever prove me right; a single experiment can prove me wrong." This is a paraphrase from his "Induction and Deduction," *Collected Papers of Albert Einstein* Vol. 7, Document 28. The Berlin Years: Writings, 1918–1921, eds. A. Einstein; M. Janssen, R. Schulmann, et al..

22. Einstein's letter to Solovine (March 30, 1952).

That is why functionality is a basic notion in the life sciences, making biologists look for the function of any biological feature—and if they cannot find one, they will and should keep searching until they do. Functionality of causes that have successful effects is as basic to the life sciences as causality is to all the natural sciences.

The next issue would be the enigma of this universe being *intelligible*. Where could this notion of intelligibility come from? It certainly does not come from science itself, of course. Scientists assume that, in principle, the world can be known and taken as intelligible—otherwise there would be no need to pursue science. So intelligibility is definitely not the outcome of intense and extensive scientific research; it is not intra- but extra-scientific; it is in fact a proto-scientific notion that must come first before science can even get started. If you were told scientists had discovered that certain physical phenomena are not intelligible, you would, or at least should, tell them to keep searching and come up with a better hypothesis or theory—based on this fundamental philosophical assumption that says the universe is "fundamentally" intelligible and comprehensible. Albert Einstein used to say that the most incomprehensible thing about the universe is that it is comprehensible, actually a mystery.[23]

I see only one reasonable explanation for this intelligibility: All human beings, including scientists, do uphold the conviction—consciously or subconsciously—that there is an intelligible plan behind this universe, a plan that is accessible to the human intellect through the natural light of reason. This intelligibility may be an incomprehensible thing to scientists, a mystery even, but not so to religious believers, for they know *why* we can comprehend this universe—because this universe was created by a rational Creator God. Without God, our universe would indeed be incomprehensible, inexplicable, not to say inexistent. Let me quote Albert Einstein again: "Every one who is seriously involved in the pursuit of science becomes convinced that a Spirit is manifest in the laws of the universe—a Spirit vastly superior to that of man, and one in the face of which we with our modest powers must feel humble."[24]

---

23. Albert Einstein, "Physics and Reality," *Out of My Later Years* (the Estate of Albert Einstein, 1956), 61.

24. Albert Einstein as cited in Dukas and Hoffmann, *Albert Einstein: The Human Side* (Princeton University Press, 1979), 33.

To sum up, at any moment in their research, scientists must *assume* that our universe can be known and made intelligible, that there is orderliness so that like causes have like effects, and that there is teleology in nature so that some designs work better than others. These are not just rules that regulate what is good science and what is bad science, but they determine what constitutes science and what does not. I hate to call them rules, since they are so fundamental that whoever violates them is not talking science anymore and is no longer a serious teammate in the scientific community.

To put it in more technical terms, the notions of order, functionality, and intelligibility are *proto*-scientific. Science cannot explain them but needs them for its explanations—but because they are so fundamental, they are easily overlooked. Yet we need to face the question: Where do they come from then? I do not believe we can just "read" them into the world around us, à la Kant. They are not an *a priori* form of thought but a given in reality. I see no better answer than that they were "engraved" in the universe, making the universe look more like "a great thought than a great machine,"[25] in the words of the late astrophysicist Sir James Jeans.

A closely related issue is the fact that we are *rational* beings. We found out that rationality does not come from the animal world, let alone through genes. So where then would our rationality come from? Gradually I had to come to the conclusion that, without God, our rationality would be of a very dubious nature. If rationality were a product of natural selection, we would have no reason to trust our reasoning. Even Charles Darwin realized this vaguely when he wondered in his Autobiography "whether the mind of man, which has, as I fully believe, been developed from a mind as low as that possessed by the lowest animal, [can] be trusted when it draws such grand conclusions."[26] And in one of his letters he raises a similar question, "Would anyone trust in the convictions of a monkey's mind, if there are any convictions in such a mind?"[27]

Darwin would have been right had he said that our convictions cannot be trusted if they came from a *brain* that developed from a monkey's brain—but instead he erroneously used the word "mind." Darwin did not

25. James Jeans, *The Mysterious Universe* (Cambridge University Press, 1930), Chapter 5.

26. Charles Darwin, *The Autobiography of Charles Darwin*, ed. Nora Barlow (New York: Norton &and Company, 1958), 149.

27. Letter to W. Graham, 1881.

seem to acknowledge that the human mind is something different from the brain—something "more" than the human brain. Hence, he did not realize that the theory of natural selection must *assume* the human mind, but can neither create nor explain it.[28] If the human mind were the mere product of natural selection, so would be science, for nothing we claim to know could then be trusted.

It is clear to me now that rationality cannot receive its power from evolution. It is equally clear that rationality cannot be defended by using rational means, for that would amount to begging the question; rationality cannot possibly establish its own rationality. Nor can it be founded on science, for it actually enables science. So where does rationality come from? Again, I only see one possible answer: It comes from creation, as a reflection of the Creator's Mind. Only then can we explain how the rationality as present in our minds corresponds with the rationality present in the world. A rational answer would be that their congruence can be best explained at a deeper, or rather higher, level, the Rationality of a Creator who respects reason. Our mind did not invent God, but the situation is actually reversed—without God, our mind could not even exist, let alone know what it pretends to know.

It is thanks to our Creator that we have a rational lawgiver who guarantees order, stability, and predictability, for it is to the essence of rationality that there are truths built into us and into the world that reason can apprehend. Truth and untruth are not just a matter of opinion. Reason is not a product of ir-rationality but it comes forth from the Great Intelligence that is behind everything. This gives us the ultimate assurance that the cosmos[29] is possessed of an inner logic accessible to human reasoning.

The physicist, mathematician, and Jesuit Roger Joseph Boskovitch (1711-1787), who developed the first coherent description of atomic theory,[30] put it very well: "Regarding the nature of the Divine Creator, my the-

---

28. Curiously enough, Charles Darwin applied this to one's belief in God (theism) but not to his own belief in evolution (evolutionism). His conclusion was that we cannot trust anything we know about God, whereas I would rather argue the opposite—that we cannot trust anything we know at all if there were no God.

29. The Greek word "cosmos" means "order."

30. His atomic theory, given as a clear, precisely-formulated system utilizing principles of Newtonian mechanics, inspired Michael Faraday to develop his field theory for

ory is extraordinarily illuminating, and the result from it is a necessity to recognize Him."[31] Indeed, without God, we would have no reason to trust our own reasoning. Without God, I would almost literally lose my mind. Nothing that I think is true and nothing that I think is right can be trusted if there is no God. Albert Einstein put it this way: "That deeply emotional conviction of the presence of a superior reasoning power, which is revealed in the incomprehensible universe, forms my idea of God."[32]

One more related issue would be the fact that we are *moral* beings. We discussed earlier that morality does not come from the animal world and is not etched in our genes. Yet, morality tells us what we ought to do for others and ought to receive from others. It gives us moral rights and moral duties; it gives us moral responsibilities that reinforce what "by nature" we would not desire to do anyway.

How could morality be such a demanding issue—demanding an absolute authority—if it were only a matter of genes, or tradition, or majority votes, or political correctness? Do my genes, or any other natural factors, have the right to demand absolute obedience from me? Of course not! Does society have the right to demand my absolute obedience? Certainly not! Does any person, including myself, have the right to demand my absolute obedience? None of the above! The only authority that can obligate me is something—or rather someone—infinitely superior to me; no one else has the right to demand my absolute obedience. And we can easily extend this to the Ten Commandments: They are God's prescription for happiness—two tablets a day. Let no one tell you that they have become irrelevant in our modern society, given the fact that few of our neighbors possess an ox or an ass for us to covet. It does not take much intellectual effort to modernize ox and ass into a Jaguar and a luxury yacht.

Even an atheist such as the French philosopher Jean Paul Sartre realized that there can be no absolute and objective standards of right and wrong, if there is no eternal Heaven that would make moral values objective and

---

electromagnetic interaction. Other nineteenth century physicists, such as William Rowan Hamilton and Lord Kelvin, stressed the theoretical advantages of the Boškovićian atom over rigid atoms.

31. Roger J. Boskovitch, *A Theory of Natural Philosophy* (Zagreb, 1974), 539.

32. In Alice Calaprice, *The Quotable Einstein* (Princeton, New Jersey: Princeton University Press, 1996).

universal. As he puts it, it is "extremely embarrassing that God does not exist, for there disappears with Him all possibility of finding values in an intelligible heaven. There can no longer be any good *a priori,* since there is no infinite and perfect consciousness to think it."[33] We would become animals without morality.

The German philosopher Friedrich Nietzsche was another atheist who also understood how devastating the decline of religion is to the morality of society, when he wrote "God is dead: but as the human race is constituted, there will perhaps be caves for millenniums yet, in which people will show his shadow [...] All of us are his murderers."[34] Nietzsche is saying here that humanism and other "moral" ideologies shelter themselves in caves and venerate shadows of the God they once believed in; they are holding on to something they cannot provide themselves, mere shadows of the past. They are "idols" constructed to preserve the essence of morality without the substance. In a world without divine and eternal laws, neither our dignity nor our rationality would be able to survive. Even the non-religious philosopher Jürgen Habermas expressed as his conviction that the ideas of freedom and social co-existence are based on the Jewish notion of justice and the Christian ethics of love. As he puts it, "Up to this very day there is no alternative to it."[35]

Because of all the above, we must recognize that morality comes from Heaven. Moral values reside in Heaven. We ought to do what we ought to do—for Heaven's sake! The *United States Declaration of Independence* makes it very clear that we are endowed by our Creator with certain unalienable Rights—not man-made but *God*-given rights, that is. When in 1948 the United Nations (UN) affirmed in the *Universal Declaration of Human Rights* that, "all human beings are born free and equal in dignity and rights," it must have assumed the same—otherwise all those rights would be sitting on quicksand, subject to the mercy of law makers and majority votes.

---

33. Jean-Paul Sartre, *Existentialism from Dostoevsky to Sartre*, ed. Walter Kaufmann (New American Library, 1975), 353.
34. Friedrich Nietzsche, *The Gay Science*, translated by Walter Kaufmann (Vintage Books 1974), § 108 and § 125.
35. Jürgen Habermas, Ciaran Cronin, and Max Pensky, *Time of Transitions* (Cambridge: Polity Press, 2006), 150-151. See also Eduardo Mendieta, ed., *Religion and Rationality: Essays on Reason, God, and Modernity* (Cambridge, MA: MIT Press, 2002), 149.

Without God, we would have no *right* to claim any rights. If there were no God, we could not claim any of those rights we think we have the right to claim; instead we would only have (legal) entitlements, but no (moral) rights. John F. Kennedy put it well in his Inaugural Address: "the rights of man come not from the generosity of the state, but from the hand of God." In response to Immanuel Kant, who said we should all start acting in a way that is moral "even if God does not exist,"[36] Pope Benedict XVI argues that we should do the opposite and live a moral life "as if God existed."[37]

Some people might object that morality cannot be based on God. In contrast, I would argue that those who say God cannot be the basis of morality because there are so many different religions commit the same fallacy as those who say evolution cannot be the explanation of the diversity of life because there are several theories as to how life unfolded. Instead I would maintain that, without God, anything is permissible—or at least we can make anything permissible by autonomously changing absolute moral values into our own relative moral evaluations. Right and wrong is not just a matter of opinion.

So we must conclude that morality cannot be ruled by science; actually it is reversed, science should be ruled by morality. We cannot even establish morality by pointing out how immoral it would be to reject morality, for that would already require a basic sense of morality. In addition, morality cannot be rooted in our genes; it is not the product of natural selection; it is not the result of any legislation; it is not a scientific conclusion; it is not based on anything useful or beneficial such as "the greater happiness of a greater number of people." All of these substitutes are morally irrelevant, since morality includes a new dimension that only morality has access to.

Be on your guard for moral substitutes. We cannot define moral notions in non-moral terms; the fact that something *is* a certain way does not entail that it *ought* to be that way; the fact that diseases are natural in a biological sense does not entail they are good in a moral sense. Or to put it more concisely, description does not automatically lead to prescription. The fact that the "survival of the fittest" may be natural does not mean that we should enforce it as a moral issue. And the fact that some people are richer than

36. Immanuel Kant, *The Metaphysics of Morals*, 80. 3.5.

37. Joseph Ratzinger, lecture given in the convent of Saint Scholastica in Subiaco, Italy, on April 1, 2005.

others or more intelligent than others does not mean that we ought to value them differently in a moral sense. Those people may have more power than others, but they do not have more rights than others. The fact that human beings *are* different does not mean they *ought* to be treated in a different way. Hence, a moral property such as being good or right cannot be reduced to a natural property such as being natural, functional, genetic, more evolved, more developed, better for the majority, or whatever.[38]

So, in my view, there is only one possibility left: Morality comes from the Great Unknown; it is written in our hearts and minds, guiding us to make the right moral choices in life. The only authority that can obligate me is something—or rather someone—infinitely superior to me; no one else has the right to demand my absolute obedience in matters of human dignity, human freedom, and the like. Moral rights and duties are absolute, objective standards of human behavior—they are nonnegotiable. We are responsible for the moral choices we make in life. Yes, we do have a choice when it comes to morality, but that does not mean we can just pick whatever we want. We cannot just vote to decide whether we are anti-slavery and anti-abortion, or not. Abraham Lincoln put it well when he challenged the Nebraska bill of 1820 that would let residents vote to decide if slavery would be legal in their territory: "God did not place good and evil before man, telling him to make his choice." There is no "pro-choice" in morality.

We were created with a moral compass, pointing not to the magnetic North but to the "Above"—to a place where justice reigns and moral values reside. "Above" is the world of what is "unseen" and "invisible." There is no other way we can trust our moral values and rights—the reason being that they are *God*-given, and hence non-negotiable. Without an eternal Heaven, there could be no absolute or objective standards of right and wrong. If these did not come from God, people could take them away anytime—which they certainly have tried to do many times.

---

38. The late Cambridge philosopher G. E. Moore calls this the "naturalistic fallacy" (*Principia Ethica*, 1903). This fallacy consists in erroneously defining moral notions in non-moral terms. The property of "goodness" cannot be defined. It can only be shown and grasped. Any attempt to define it (X is good if it has property Y) will simply shift the problem (Why is Y-ness good in the first place?).

Let me put all these rational and reasonable "arguments" together in one long argument:

- Could there be design in nature, if there were no intelligent Designer?
- Could nature be intelligible if it were not created by an intelligent Creator?
- Could there be order in this world if there were no orderly Creator?
- Could there be scientific laws if there were no rational Lawgiver?
- Could there be moral laws if there were no moral Lawgiver?

I have come to the conclusion that the answer to all these questions is a rather definite "No, there could *not*..." Through the way the universe is, we are able to know quite a bit about the Great Unknown "beyond" and "behind" the universe: He must be intelligent, orderly, rational, and moral. Is this an infallible, conclusive proof of a Creator God? No, it is not, but it certainly is a powerful pointer to a trustworthy Creator. How so? Once we lose the notion of creation, plus the trustworthy order it comes with, natural laws and moral laws would be resting on quicksand.

To put this in more general terms, our universe is best understood as an "intelligent project."[39] The Nobel Laureate and physicist Arno Penzias worded this as follows: "Astronomy leads us to a unique event, a universe which was created out of nothing, one with the very delicate balance needed to provide exactly the conditions required to permit life, and one which has an underlying (one might say 'supernatural') plan."[40] Even the physicist Paul Davies said at one point, "There must be an unchanging rational ground in which the logical, orderly nature of the universe is rooted."[41] The cosmic design must be an intelligent design.

As a matter of fact, without God, we would have no rational explanation

---

39. Do not confuse this with the *Intelligent Design Theory* that some critics of evolutionary theory have advanced.

40. Arno Penzias as quoted in H. Margenau and R. A. Varghese, ed. 1992. *Cosmos, Bios, and Theos* (La Salle, IL: Open Court), 83.

41. Paul Davies, "What Happened before the Big Bang," *God for the 21st Century*, edited by Russell Stannard (Templeton Press, 2000), 10-12. Later on, however, Davies changed his opinions (e.g. in "Taking Science on Faith," *New York Times*, 2007-11-24).

as to why the universe is the way it is. Due to the framework of an intelligent design behind this universe, we as human beings have been able to emerge as intelligent designers ourselves, capable of designing structures guided by art, architecture, engineering, and scientific research. If my mind can go up "into the sky" to take a mental bird's-eye view of my world, of my life, of my activities, why would it not be able also to go up "into heaven" to see everything from an even higher perspective—God's point of view, if you will? Were we not made in his image, as it is?

Denying that there is a Creator God would actually be an acid eating away the foundation of rationality and morality. The reason to trust our reasoning as well as the right to claim our moral rights can only be found in someone "above" us, God. One could even say that man can be understood only in light of God. Denying the existence of God would even destroy and undermine science: Science would become impossible. Apparently, even science is a faith-based enterprise. It is only because of their trusting that nature is law-abiding and comprehensible in principle that scientists have reason to trust their reasoning.

Curiously enough, as far as science is concerned, there is much more believing in what we know than many want to believe. And, vice-versa, in religion, there may be much more knowing in what we believe than many seem to know. Let me quote the famous physicist, Nobel Laureate, and founder of quantum physics Max Planck when he said, "Both religion and natural science require a belief in God for their activities; to the former He is the starting point, and to the latter the goal of every thought process."[42] Planck would certainly be joined by the famous French chemist Louis Pasteur, founder of medical microbiology, who said "Science brings men nearer to God."[43]

It should not surprise us then that the Judeo-Christian concept of a Creator God is the cornerstone of all of the above. The Judeo-Christian belief is that nature is not a divine but a *created* entity; nature is not divine in itself, only its Maker is—which opens the way for scientific exploration

---

42. Max Planck, "Religion and Natural Science" (1937) in Max Planck and Frank Gaynor (trans.), *Scientific Autobiography and Other Papers* (New York: Philosophical Library, 1949): 184.

43. J. H. Tiner, *Louis Pasteur—Founder of Modern Medicine* (Fenton, MI: Mott Media, 1990), 90.

(otherwise we would not be allowed to "touch" the divine). Because the Judeo-Christian God is a reliable God—not confined inside the Aristotelian box, not capricious like the Olympians in ancient Greece, and not entirely beyond human comprehension like in Islam—the world depends on the laws that God has laid down in creation. Thanks to God's creation, everything else has lost its power, has lost its divine allure. Faith in the one God changes the universe, once inhabited with spirits, deities, and goddesses, into something "rational." Only God is the source where the order as well as the intelligibility of the universe ultimately stems from.

The only way to find out what this order looks like is to "interrogate" the universe by investigation, exploration, and experiment. The door is wide open. Through scientific experiments we can "read" God's mind, so to speak. So, not surprisingly, God has often been described as an Architect, an Artisan, or a Workman (*Deus Faber*). Relating to the Book of Wisdom, we could even say that the created world receives weight, number, and measure through the hands of God.[44] That is the reason why the Nobel Laureate and physicist Charles Townes, who invented the laser, could say, "When God said 'Let there be light,' he surely must have meant perfectly coherent light."[45]

This religion-inspired thought has a long history in science. Copernicus' achievements were in fact based on his religious belief that nothing was easier for God than to have the earth move, if he so wished. And Kepler's Christian belief told him God would not tolerate the inaccuracy of circular models of planetary movements in astronomy. It was in God that these scientists found reason to investigate nature and trust their own scientific reasoning. The founder of quantum physics, Max Planck, put it well: "It was not by accident that the greatest thinkers of all ages were deeply religious souls."[46] People who come to mind are Johannes Kepler, Isaac Newton, Blaise Pascal, Gregor Mendel, Louis Pasteur, George Lemaître, and so many more.

In short, it is the Judeo-Christian view that the universe is the creation

---

44. Book of Wisdom 11:20.

45. Charles Townes, "Laser Physics: Quantum Controversy in Action," in Benjamin Bederson (ed.), *More Things in Heaven and Earth: a Celebration of Physics at the Millennium* (New York: Springer, 1999), 443.

46. Max Planck, *Where is Science Going?* (Brooklyn, NY: AMS Press, 1977), 168).

of a rational Intellect that is capable of being rationally interrogated by all human beings, including the scientists among them. I believe it is quite evident that nature remains an enigma until nature is no longer ruled by whimsical deities, chaotic powers, or our own philosophical decrees and regulations. Whether they like it or not, all scientists keep living off Judeo-Christian capital. Even the late nuclear physicist J. Robert Oppenheimer, who was not a Christian, was ready to acknowledge this very fact when he said, "Christianity was needed to give birth to modern science."[47] Unfortunately, many scientists are unaware of this fact. When you think that what you have is all there is, you do not know what you are missing. Those who live in ignorance do not know they do; if they were able to know that, they would no longer be ignorant.

As a matter of fact, science was born in the cradle of Judeo-Christian faith, especially during the Late Middle Ages. Here are some of its pioneer crafters: At the end of the first millennium, Pope Sylvester II had already used advanced instruments of astronomical observation, in his passion for understanding the order of the universe. Albert the Great (or Albertus Magnus, 1206–1280), the teacher of Thomas Aquinas, had quite a track record for his time, when he discovered the element arsenic; experimented with photosensitive chemicals, including silver nitrate; and made disciplined observations in plant anatomy and animal embryology. During the same time period, Bishop Robert Grosseteste introduced the scientific method, including the concept of falsification, while the Franciscan friar Roger Bacon established concepts such as hypothesis, experimentation, and verification. In other words, what some consider a period of darkness—during the so-called "dark" Middle Ages—was actually the birth of the light of reason.[48]

So I ask the question again: Is there someone out there? Yes, I believe we can say there is, but not somewhere out in space, for such a power would be as limited as our own power, even if raised to the zillionth power. No,

---

47. J. Robert Oppenheimer, "On Science and Culture," *Encounter* 19, 4 (1962): 3–10.

48. It always amazes me how someone like Carl Sagan, in his book *Cosmos, One Voice in the Cosmic Fugue* (New York: Ballantine Books, 1980), makes it look as if nothing happened in science between 415 AD and 1543 AD. That is just historical ignorance. For a good analysis of this issue, see James Hannam, *God's Philosophers: How the Medieval World Laid the Foundations of Modern Science* (London: Icon Books, 2010).

it is not someone in outer-space, but someone beyond the "outer-space," someone without space and time, with unlimited power, with power of a different degree and nature—a transcendent power, that is. Only such a kind of power would be capable of being the Maker of Heaven and Earth.

## Laws and Purposes

Our universe is a law-abiding universe—a universe of law-and-order. That is one of our fundamental beliefs—even in science. But as I argued in the previous chapter, this belief is ultimately based on a trustworthy Creator God who created a "reliable" universe—not a haunted house or a bizarre fairy tale; we can orient ourselves in it, feel secure in it, and make plans for its and our future. Everything seems to be a matter of cause-and-effect, with like causes having like effects. Do not think that science has demonstrated this experimentally, for if we did find that like causes produced unlike effects, we would automatically assume that those causes must have somehow been different or that some hidden causes must have interfered.

So the next question would be: If there is a God, how can God still work in such a law-abiding universe? If God is the Primary Cause, can he do more than let the secondary causes do their work "in obedience to" his laws of nature? Do the laws of nature not take care of everything—from planetary movements to evolutionary developments to genetic processes? It seems so, does it not? Yes, in a way it does. There is no need for the Primary Cause to interfere with secondary laws, since those do their own job. If God, the Primary Cause, had to interfere, he would become a secondary cause that works *inside* the universe in order to fill the gaps left behind by all the other secondary causes. That would degrade God.

Yet, even the famous physicist Isaac Newton did fall for this timeless temptation of having God keep a "divine foot" in the door, when he called upon God's active intervention to periodically reform the solar system from increasing irregularities, and to prevent the stars from falling in on each other, and perhaps even in preventing the amount of motion in the universe from decaying due to viscosity and friction. Today, we know God does not have to make these interventions, because science can now explain them with the proper laws (which are God's laws anyway). Newton wrongfully made the Primary Cause periodically act like a secondary cause.

Even nowadays, proponents of the Intelligent Design Theory are doing something similar in evolutionary biology: Their theory claims that certain features of the universe and of living things are best explained by an intelligent, supernatural cause, in addition to or in replacement of natural processes such as mutation and natural selection. Again, what they are doing is making the Primary Cause periodically act like a secondary cause working *inside* the universe. In doing so, they change God into a god-of-the-gaps—a god who fills in the gaps left behind in his allegedly defective design of the universe. The great scholastic theologian Francisco Suárez put it this way: "God does not intervene directly in the natural order where secondary causes suffice to produce the intended effect."[49]

In contrast, I would claim that God did not outfit the universe with a defective, incomplete design. Moreover, God should never be a "missing link" in our scientific explanations, nor should the Creator be an element or part of his own creation working from *inside* the universe. This kind of god often proves to be a fleeting illusion, for when the frontiers of science are being pushed back—and they usually are—this kind of god would be pushed back with them as well.

Instead, I would argue that God is the very origin of all that science discovers; he works from *outside* the universe and thus permeates everything on the inside, but he should never become a scientific element within his own creation. Please, let us avoid that pitfall. In the words of the Harvard astronomer and historian of science Owen Gingerich, "I.D. [intelligent design] is interesting as a philosophical idea, but it does not replace the scientific explanations that evolution offers."[50]

In another way, though, the laws of nature may *not* take care of everything. What is missing then? If everything were solely a matter of causes and effects, there would be no room left for something like purposes. Yet, I firmly believe there should be room left for *purposes*. I must admit, though, purposes do not seem to fit in a cause-and-effect world. As a matter of fact, scientists have rightly decided that "purposes" should be removed from the scientific picture as non-measurable, non-quantifiable, non-mathematical, non-mechanical, and non-physical entities.

---

49. Francisco Suárez, *De Opere Sex Dierum*, II, c. x, n. 13.

50. Owen Gingerich, "Taking the ID debate out of pundits' playbooks," *Science & Theology News*, Nov. 8, 2005.

As a matter of fact, the concept of "purpose" was taken out of astronomy by Nicolas Copernicus, out of physics by Isaac Newton, and out of biology by Charles Darwin. Astronomers do not seek the purpose of comets or supernovas, nor do chemists search for the purpose of hydrogen bonds. The concept of purpose plays no part in scientific explanations. The sun does not rise every morning because it "wants" to, but because it follows physical laws. Water "seeks" its own level, but it does not do so intentionally. Even plants do not "seek" the light, but they do respond to the light in their environment through a light-sensing hormone, called auxin, which makes plant shoots curve towards light. Eye patterns on butterfly wings have the effect of warning enemies. That is a function of eye patterns, not a purpose of butterflies. In all such cases, "purposes" have been replaced with causal mechanisms.

So do I want to bring purposes back in? Yes and no. Part of my answer is no. Purposes clearly do not belong in science because they are not measurable, quantifiable, material, or physical. But once we acknowledge this viewpoint, we should stick with our own decision to discard purposes from science, and accept its consequences. Once you remove purposes from scientific discourse, you cannot keep referring to them. Nonetheless, some scientists breach their own decision as they continue talking about purposes, or the lack thereof.

Here is just a tiny selection of those inconsistent scientists. The chemist Peter Atkins tells us, "We are children of chaos. [...] Gone is purpose; all that is left is direction. This is the bleakness we have to accept as we peer deeply and dispassionately into the heart of the Universe."[51] Or take this one from the biologist Douglas Futuyama: "We need not invoke, nor can we find any evidence for, any design, goal, or purpose anywhere in the natural world."[52] Perhaps the most arrogant one comes from the late paleontologist George Gaylord Simpson, who ventured to proclaim from his quasi-scientific pulpit, that "man is the result of a purposeless and natural process that did not have him in mind."[53]

---

51. Peter Atkins, *The Second Law* (Scientific American Library, 1984), 200. In the last paragraph of chapter 9, he seems to express himself more carefully.

52. Douglas Futuyma, *Evolution* (Sunderland, MA: Sinauer Associates, 2005), 12.

53. George G. Simpson, *The Meaning of Evolution* (New Haven: Yale University Press, 1967), 345.

My problem with these scientists is that one cannot make any of these statements as a scientist. First they eliminate the concept of purpose from science, but then they keep using it by rejecting it. Once "purpose" has been eliminated from science, it can no longer be used, let alone be explained, by science, as it is forever beyond science's reach. Darwinism, for instance, just does not know whether evolution has a purpose or not, for the simple reason that the word "purpose" does not exist in its vocabulary.

Yet, part of my answer to the question "Do I want to bring purposes back into the discussion?" is also a definitive yes. When scientists removed "purposes" from scientific discourse, they did not make them disappear entirely, of course; they just moved them from inside to outside the scientific domain. The fact that "purposes" are missing on scientific maps does not mean they do not exist at all; they are not completely out of the picture, although they are out of the scientific picture. Remember, the fact that human beings are missing on astronomical or geological maps, does not entitle us to deny their existence; whatever we neglect we can never just reject.

Besides, maps tell us *how* to go, not *where* to go; destinations and purposes do not occur on maps but they do reside in the minds of map-makers and map-users. As we all know, the direction of something like a billiard ball on the pool table is not only ruled by physical laws but also by purposes that players have in mind. Human beings have the ability to steer the laws of nature in a specific direction, according to certain purposes residing in their minds. It is those purposes that can become causes of actions on their own, so they are definitely possible, even in a world of cause-and effect.

It is time now to ask some pertinent philosophical questions for those who keep denying the existence of purposes. If there is no purpose in the universe at all, how were we ever to know there is no such thing as a purpose? As C. S. Lewis put it, "If there were no light in the universe and therefore no creatures with eyes, we would never know it was dark."[54] Besides, I would like to ask those who deny the existence of purposes what the purpose is of trying to prove that there is no purpose in life. As a matter of fact, denying that there are purposes in life defeats its own claim. If it is your purpose to remove all purposes from life, you are also wiping out

54. C. S. Lewis, *Mere Christianity* (San Francisco: Harper, 2001), 46.

your own purpose of doing so; those whose purpose it is to eradicate all purposes from life have lost even the very purpose for doing so.

That is why it always strikes me how some people make it their main purpose in life to claim that there are no purposes in life at all. I think the British philosopher and former atheist Anthony Flew worded it well when he said, "How can a universe of mindless matter produce beings with intrinsic ends, self-replication capabilities, and 'coded chemistry'? Here we are not dealing with biology, but an entirely different category of problem."[55]

I like to take this thought one step farther: In a universe without purposes, there could not even be any man-made machines, since such machines, curiously enough, are always made for a purpose; the world of technology is per definition purpose-driven, based on purposes that designers and engineers have in mind. Therefore, we could never ban purposes from the universe by saying the universe is just a machine that runs with clockwork precision. Using the machine metaphor to claim there is no purpose in this universe is a bit odd, to say the least.

Let us acknowledge that issues of sense, meaning, and purpose do not belong in scientific discourse—they cannot feature on scientific maps—and yet they do exist in the minds of map-makers and map-users. I say it again: Science may be everywhere, but science is not all there is. The question as to whether evolution has a meaning, destination, or purpose takes us into the domains of philosophy and religion. Science just has no answers to these kinds of questions—no matter what some scientists proclaim, or like to proclaim. So scientists have no right as scientists to decree that we are unintended, fortuitous creatures or mere products of a blind and purposeless fate. That is beyond their competence as scientists.

Yet it is a longstanding fallacy of making such claims. Hume, for one, happily fell for it: "But the life of a man is of no greater importance to the universe than that of an oyster."[56] The fact that we are contingent does not entail we are "accidental" in the sense of mere products of blind fate. Do not fall for the temptation to capitalize the word "chance" by changing it

---

55. Antony Flew, *There is a God: How the World's Most Notorious Atheist Changed His Mind* (New York: Harper One, 2007), 124.

56. David Hume, "On Suicide" (written 1755, published 1777), in Stephen Copley and Andrew Edgal (eds.) *David Hume: Selected Essays* (Oxford University Press, 2008), 319.

into the goddess of Fate or Doom or Blind Fate. Science has nothing to say about chance with a capital C. Fate is far beyond its reach, for it is in essence a worldview notion. When we speak of randomness in science, we are talking in statistical terms, in the sense of how things in this universe are related to each other—not how they are related to God.

At the very moment we take chance as a fate issue—in the sense of "being meaningless, unintended, and purposeless"—we have left the territory of science. In the words of the physicist Stephen Barr, "one must distinguish between words used by scientists and words used scientifically."[57] No matter how scientists believe evolution operates, chance with a capital C is absolutely beyond their reach and expertise. Yet, outside the scientific domain, there is still ample space left for purposes. As the physicist and theologian Ian Barbour puts it, "The goal of science is understanding lawful relations among natural phenomena. Religion is a way of life within a larger framework of meaning."[58]

That having been said, we should also acknowledge that purposes do not belong to the visible, material world but to the invisible, spiritual world—that is, to the same immaterial world where we discovered natural laws and objective truths, as well as moral laws and moral values. Here we are dealing with a world without which we cannot live as moral and rational beings. A world-view claiming there are only material things is as fragile as the "material" that generated this world-view. Human beings always have purposes in mind. Take those away from them, and they are mere animals—arguably back to where they came from.

In general, one might say that causes rank high in science, whereas purposes rank high in religion. Are they in conflict with each other? I do not see how they could possibly contradict each other. It must be clear by now that I strongly advocate a "separation" of science and religion, with a good fence separating them and each one having its own authority and expertise. Why do I consider this to be the best policy?

Some people like to mix them together into one concoction. They tend to speak in terms of "reconciliation," as if science and religion were just

57. Stephen M. Barr, "Chance, by Design," *First Things*, 2012, 12, 26.

58. Ian G. Barbour, *Religion and Science* (HarperOne, 1997), 204.

two different ways of expressing single truth. Such people basically want to "reconcile" the data of religion with the data of science. They want to fuse two different perspectives into one—a grand synthesis of science and religion, so to speak, merging them into the kind of hodgepodge someone like the Jesuit and paleontologist Teilhard de Chardin tried to brew. However, if you mix them together, it can go into two different directions: Either science annexes religion or religion takes over science. What is wrong with that?

The most common outcome of a fusion is that science annexes religion—"interpreting" the data of religion, that is, so as to leave the data of science intact—the reason for this being that scientific facts are considered to be safe and "proven." In contrast, we found out that what we call "proven" scientific knowledge is only proven until a new set of empirical data "disproves" what was previously considered "proven." Besides, there is no reason why religion should bend to the criteria of science. What gives science that authority?

The second possible outcome of fusion is that religion conquers and suppresses science by transforming scientific data with a religious touch. But I consider such a route dangerous as well; it has been tried out many times—most recently by proponents of the "intelligent design theory." They invoke supernatural causes in addition to natural causes; they declare science basically incompetent and incomplete, and the *Book of Nature* inconsistent and incoherent. Like all other attempts of merging science with religion into one concoction, this attempt tries to create a mixture of "oil and water"; but these two just do not mix well together.

Instead, I would highly recommend respecting their distinctive powers and their different angles of approach. Call it a "separation" of science and religion, if you want—analogous to the separation of church and state. They are both autonomous, which is different from being autocratic and authoritarian, for there is always "another side to the story." If we do not honor their separation, science will seek divine authority and religion will claim scientific authority.

I admit that when you try to keep them separate, you may run the risk of placing them in competition with each other, having religion silenced by science (or science silenced by religion). One of the arguments used to do so is that science at least shows progress, whereas religion seems to cling to its dogmatic, eternal truths. This suggests that if one religion be true, then all must be true—which basically means they all must be *false*. This puts the validity of any kind of religious perspective in contrast to a scientific approach. Someone like George Bernard Shaw took this side when he said that religion is always right whereas science is always wrong—and that's why he preferred science over religion.[59] Well, I have news for Shaw: He was wrong himself. In line with his comparison, he could have replaced religion with mathematics: Mathematics is always right whereas science is always wrong. Would this really force us to prefer science over mathematics? We are comparing apples with oranges here.

In defiance of what Shaw said, I would point out that progress in science only means that scientists have come to "see" more and more things in a better and better light—thanks to a more and more refined set of concepts that remains constantly open to revision. Yet, in science, there is much more "believing" in our "knowing" than many want to believe, as we found out earlier—so why could there not be much more "knowing" in religion, in "believing," than many seem to know? When comparing science with religion, we should acknowledge that they both have a component of "knowing" as well as a component of "believing," but the former one is stronger in science, whereas the latter one is stronger in religion.

Unfortunately for religion, though, the power of strict proof grows to be weaker when the field of reference becomes wider. Logic and math, for example, have a strong power of proof but happen to be about very little. Religion, on the other hand, has the reversed "problem": There is so much in life that counts, and yet cannot be counted or measured. Religion and faith are certainly not "blind" for reality but try to make the best sense of everything there is on the basis of the limited evidence available.

So I think we should come to the conclusion that there is a legitimate task, even a need, for religion—no matter how strong science has become and how compelling its advances seem to be. Render to science what is material, measurable, and visible, but do not render to science what is God's. Without religion, science would have no foot to stand on,

---

59. Speech at the Einstein Dinner, Savoy Hotel, London (28 Oct 1930). Reproduced in George Bernard Shaw and Warren Sylvester Smith (ed.), *The Religious Speeches of George Bernard Shaw* (1963), 83.

and without science, religion would lose its contact with reality. Religion should not be steered by science—if anything, religion could steer science instead. Science should protect religion from errors, and religion should keep science within rational and moral bounds.

Here is another way of saying this: Science is in search of causes, religion is in search of purposes. We need them both, for they have different answers to very different questions. Science, on its own, is knowledge and understanding in a very narrow and specific sense; in order to reach its fullness, it needs to be broadened and expanded toward a higher truth.

If you can go along with all of this, you have entered a completely new world with a wide open horizon. My personal experience is, to paraphrase Augustine, that God has always been with me, but I was not with God for a long time.

Take the following vista: If human beings can have purposes, why could not God have purposes as well? After all, we were created in God's image! If *we* can steer our bodies with purposes that we have in mind, why could *God* not do something similar by steering the world with purposes that he "has in mind"? If human beings can have plans for their lives, why could not God have plans for his world? As I said earlier, if we would only focus on the physical laws behind a golf game, we would miss out on what the game is all about. If we would only focus on the physical laws behind this universe, we would also be missing out on what God's creation is all about.

In other words, God could very well deal with his creation in a manner similar to the way we deal with the games we play, the designs we construct, and the research we conduct.[60] Does this mean we go *against* the laws of nature when we do this? No, it does not. But we can go *beyond* the laws of nature, and thus use them for the purposes we have in mind. If this is true for human beings, then God can also go beyond his own laws to deal with this world, without going against his own laws. In short, God may very well have plans and purposes "in mind" for us, even in a universe of "law-and-order."

60. We try to understand here how God provides through the way we steer and design things ourselves. But the issue is actually reversed: We are able to steer and design things because we were made in God's image.

## A Purpose-Driven World

I hope you can share my observation that purposes are in fact the driving force behind our human world. If so, we should note that those purposes are very interesting entities, to say the least. Animals do not have purposes. The green color of a caterpillar, for instance, has a *function*, namely, to deceive potential predators; such is their end or goal—or in more neutral, biological terms, such is their function. You could also say that camouflage is "for" deceiving, just like a knife is "for" cutting. But make no mistake, green caterpillars achieve a goal without having that goal as a "purpose in mind"; they were just born that way. So they show signs of inborn functions, but not of intended purposes. Whereas purposes may be something in the mind of a human product-maker, functions and designs are a feature of the product itself.

Since we are masters in anthropomorphic thinking, we often speak of "purpose" even when there is no mind in sight; we tend to say that the heart pumps for the purpose of circulating blood, but such is not the purpose of a heart beat but its function. I admit the question "What is the purpose?" does come naturally to those who are surrounded by tools, utensils, and machines. But try not to forget that what is appropriate for those man-made artifacts may not be appropriate for rocks, stars, or organisms.

So let us keep our terminology straight: In the biological world, the goal of a design is not a purpose in an animal's "mind"; a biological design works *as if* it is steered by a purpose—but that is anthropomorphic language. In other words, things like a heart, an eye, a DNA molecule, a hormone, etc. do not have a purpose, but they do have a goal—or actually an effect—that has been beneficial in the "struggle for survival and reproduction." Natural selection favors good designs that work well and fit well; it promotes causes with successful effects. Hormones, for instance, do in fact reach their target cells, but not so because they have a purpose that makes them go there!

Some biologists might vehemently protest when they hear me say that animals do not have purposes—and they probably have many followers behind them. Most of us think that animals do have goals, intentions, beliefs, and purposes. Even a philosopher such as Ludwig Wittgenstein once said, "What is the natural expression of an intention? Look at a cat when it stalks a bird; or a beast when it wants to escape."[61] But then another one, the late Harvard University philosopher W. V. Quine, said that when we attribute beliefs to animals we are imagining ourselves in the animal's shoes, so to speak, and saying on its behalf what we imagine we would think or be prone to say if we were barking at a cat up a tree or lunging at a toreador with our horns.[62]

Yet Wittgenstein seems right—one cannot help seeing a cat stalking a bird as in some sense expressing an intention. One reason why he seems to be right is that we are born "anthropomorphists."[63] Another reason is that Darwinism, or actually its ideological version called evolutionism, makes us believe that the difference between animals and humans is one of degree, not kind,[64] so animals must be humans-in-the-making, and humans, for their part, must be glorified animals.[65] In this view, animals are bound to have something like purposes "in mind," since we do. That is where I think we tend to go off the right track. Without any doubt, one can study humans as animals—which is methodological reductionism—but to say that humans are "nothing but" animals is an ideological form of reductionism—ontological reductionism, that is.[66] The fact that humans are animals does not entail that they are nothing more than animals.

Why do I question then that animals have intentions, beliefs, and purposes "in mind"? When speaking of intentions, we are referring to behavior that is chosen. So we should ask: Do animals truly choose—on a rational basis, that is? In order for it to have intentions, an animal must have

61. Ludwig Wittgenstein, *Philosophical Investigations*, transl. G. E. M. Anscombe (Prentice Hall, 1973), §647.

62. W. V. Quine, *Word and Object* (Cambridge, MA: M.I.T. Press, 1960), 219.

63. Donald Davidson suggests that "Attributions of intentions and beliefs to animals smack of anthropomorphism." ("Thought and Talk," In: *Mind and Language*, ed. S. Gutenplan (Oxford University Press), 7-23)

64. In his introduction to a book chapter of *The Descent of Man* (1871/1896, p. 66), Darwin stated: "My object in this chapter is to show that there is no fundamental difference between man and the higher mammals in their mental faculties."

65. For a more detailed discussion see Gerard M. Verschuuren, *Darwin's Philosophical Legacy: The Good and the Not-So-Good* (Lanham, MD: Lexington Books, 2012), chapter 18.

66. C.S. Lewis sarcastically spoke of a "trousered ape" in his 1947 book *The Abolition of Man*.

gone through some kind of deliberation and then decide not to deliberate any further unless new and relevant information comes to light. Human beings, on the other hand, are habitual deliberators; they are constantly making plans, refining old plans, evaluating their ends, etc. We can exit the physical stream of cause-and-effect for a moment and think before we act.

Can animals do the same by going through some kind of deliberation? Let me use the computer analogy again: A computer can be programmed to make decisions with an if-then-else structure. But this does not mean that the computer deliberates and decides—programmers did, and they make it look like the machine did. The "reason" why a calculator adds or a pump pumps is that a machine was designed for that purpose—and not because it "intends" to.

Do animals really go through some kind of deliberation, or is their behavior rather the outcome of a genetic program or a learning process based on reinforcement, after having been rewarded for a certain outcome in the past? The problem with most biological research done in this area is that test results taken as evidence for conditional reasoning can equally well be explained by associative conditioning or simple stimulus-response learning.[67] To be sure, animals may very well be able to predict each other's behavior, to view other beings as beings with predictable actions and relationships, and to recognize that social interactions have predictable outcomes. But does this mean they can read each other's "mind"? It may not mean more than that they have learned to associate certain behaviors with certain outcomes. This may very well be a matter of correlation, but not necessarily of causality, let alone intentionality.[68] Dogs used by the blind and by canine teams have gone through extensive conditional training so they look like humans in disguise.

Mere correlation is enough to enable prediction, though. When an antelope sees a lion approaching, it "knows" what is about to happen

---

67. Forgotten is *Ockham's razor*: The simplest of all explanations compatible with the evidence at hand should be considered the most probable. This principle was also laid down in *Morgan's canon* dating from 1893, which states that animal behavior should be explained by the simplest neural mechanisms available.

68. Correlation is necessary for causality but not sufficient. In terms of reasoning: Correlation can be expressed in if-statements; causality would be more like a matter of if-and-only-if-statements.

next, but this could very well be learned, or even inborn, behavior. It is very doubtful that a dog barking at a Siamese cat up an oak tree has any concepts or beliefs about cats and trees, let alone more specifically about Siamese cats and oak trees. It has even been shown that monkeys fail to understand the relationship between cause and effect.[69] Young animals do not run to their mother because they "think" there is "reason" to do so, but they are programmed that way; if they were not, their unfit program would be eliminated soon—natural selection takes care of that.

Consider the following case: How on earth do you learn to fear snakes? If you had a bad experience with a snake, you will probably be dead by now, but if you did not, you could not have learned that snakes are frightening. Research indicates that there seems to be an inborn factor involved, favored by natural selection. Although humans may have inborn fears, they can often overcome such fears, as snake pet owners can tell you. But we do not have an inborn fear for mosquitoes, so we must discover by reasoning and investigation that these can transmit the malaria parasite, the West Nile virus, the equine encephalitis virus, and probably much more. Making such connections is not inborn or conditioned, but requires the capacity of rationality.

Do not take me wrong, animals surely have emotions, drives, and instincts—which are of a physical nature—but do they fear snakes because they have a purpose in mind of avoiding danger? I seriously doubt whether animals have any purposes in mind at all—for the simple reason that they do not have a mind (and purposes come with a mind). Many people think that the mind is situated in the brain and that the mind may even be identical to the brain—so if animals have a brain, they are believed to have a mind (and therefore purposes). In spite of many claims to the contrary, I maintain they do not. What is it then that makes the mind so different from the brain? There are many reasons for making a distinction between the two.[70] Let me just mention a few.

---

69. See, for instance, Clive D. L. Wynne, *Do Animals Think?* (Princeton University Press, 2004). Elephants that had learned to lift a lid to retrieve food from a bucket continued to toss the lid before retrieving the reward when the lid was placed alongside the bucket while the food was simultaneously placed inside the bucket—raising the possibility that they have no understanding of this simple causal relationship, in spite of their intelligence.

70. For a more detailed discussion see Gerard M. Verschuuren, *What Makes You Tick? A*

First of all, humans have a mind which allows them to communicate with others through symbols, along with signals, whereas animals can only use signals. The brain can handle signals and images, but only the mind can deal with symbols and concepts. Bees have brains that can manage an intricate signaling system, but there are no symbols or concepts involved. Let me explain the difference between the two. If we train a dog to associate a command such as "The boss!" with its real boss, then that dog has been conditioned through training to respond to the command "The boss!" by looking for the real boss. The sound in itself has become a *signal* for the "real thing." Signals, like images, depend on the actual presence of the "real thing," due to associative conditioning (training). The dog has a physical image of its boss, but it has no mental concept of what a boss is like.

Humans may use the exclamation "The boss!" also as a warning *signal* to alert others, but in addition, they can use it as a *symbol* for a "mental object" of a boss. They often use that phrase only when the physical boss is *not* around. They may use their "mental object" of a boss to debate, for example, why their actual boss does not qualify as a boss, or which real boss they would prefer instead. The word "murder," for instance, may make you shiver, even if you had never witnessed a real murder; it is the mental concept of murder that causes this, not necessarily the actual event. Mental concepts transform "things" of the world into "objects" of knowledge, thus enabling humans to see with their "mental eyes" what no physical eyes could ever see before. As I said earlier, animals live in a world of events, but humans also inhabit a world of facts—which are mental interpretations of those events. Without concepts, there would not be any facts, because facts are the objects of our thoughts, and thoughts are a product of the mind, not the brain.

The previous paragraph suggested already that the difference between signals and symbols runs parallel with the distinction between images and concepts. Images can have some degree of generality—we can visualize a circle without imagining any specific size—but concepts, on the other hand, have a universality that images can never have—the concept circle applies to every circle without exception. This distinction sets humans and animals apart again. Animals can form general images, but they cannot

---

*New Paradigm for Neuroscience* (Antioch, CA: Solas Press, 2012).

form universal concepts. That's where human interpretation comes in. Images are inherently ambiguous, open to various interpretations; so we need concepts to give them a specific interpretation.

My second reason for making a distinction between brain and mind is that thoughts in the mind are very different from neural activities in the brain. The German philosopher Gottfried Leibniz once suggested to picture the brain so much enlarged that one could walk in it as if in a mill.[71] Inside, we would only observe movements of several parts, but never anything like a thought. For this reason, he concluded thoughts must be different from physical and material movements and parts. Nowadays, the mechanical model of cogs and wheels that Leibniz used has been replaced by the chemical model of biochemical pathways, but the outcome is the same.

If Leibniz is right, and I think he is, his analysis would explain why brain scans never reveal thoughts; all they can pick up are "brain waves," but never thoughts, since those fail to show up on pictures and scans. Because of this difference, we should sharpen our terminology: Neuroscientists are not mind-readers, neuro-surgeons are not mind-surgeons, and neuro-science is not mind-science. Simply put, thoughts are more than brain waves—in the same way as love is more than a chemical reaction. Medical professionals can read and interpret an electroencephalogram (EEG) or a magnetic resonance image (MRI), but looking at these does not show them any thoughts—perhaps memory "traces" of thoughts, but not the thoughts themselves. Neuroscientists just cannot "read" your mind. If they want to associate certain brain activities with certain mental activities, they need to ask you what you were thinking.

This reminds me of an observation that was made by the philosopher Ludwig Wittgenstein.[72] Picture yourself watching through a mirror how a scientist is studying your opened skull for "brain waves." Wittgenstein once noted correctly that the scientist is observing just one thing—outer brain activities, but the "brain-owner" is actually observing two things—the outer brain activities via the mirror as well as the inner thought processes that no one else has access to. In order for them to make the connection

71. Gottfried Leibniz (1646–1716), *Monadology* (Charlestown, SC: Forgotten Books, 2008), section 17.

72. Ludwig Wittgenstein, *The Blue and Brown Book* (New York: Harper & Row, 1980).

between "inner" mental states and "outer" neural states, scientists would depend on information that only the "brain-owner" can provide.

The world of the mind is only accessible to the "brain-owner." This is even so in court, in spite of lie detector tests; very often the only ones to know who did the crime or did not are the defenders themselves. Apparently, there is no mind-reading through brain scans. Would it not be nice if the "brain-observer" could read off from the lighted-up areas of the brain all the knowledge the "brain-owner" had gained in a life time? If this were so and we could still scan Einstein's brain and we could all become little Einsteins.

Third, thoughts are immaterial and can be true or false. Whereas the brain as a material entity has characteristics such as length, width, height, and weight, the mind does not have any of those; thoughts are true or false, right or wrong, but never tall or short, heavy or light (unless taken in a figurative sense). The late Nobel Laureate Sir John Eccles stressed the difference as follows: "The more we discover about the brain, the more clearly do we distinguish between the brain events and the mental phenomena, and the more wonderful do both the brain events and the mental phenomena become."[73]

If the mind were just the brain, its thoughts would be as fragile as the molecules they supposedly came from. In contrast, we need to stress that mental activities are very different from neural activities—we cannot just deny the mental, because denying the existence of mental activities is in itself a mental activity, and thus would lead to contradiction. Some have worded this along the following lines: If mental processes are nothing but the motions of atoms in the brain, we have no reason to suppose that our beliefs are true and hence we have no reason for supposing our minds to be composed of atoms.[74] I consider this a profound philosophical statement that you may have to digest for a while. It should shake the atoms in your head.

Fourth, the brain is only the physical carrier of immaterial thoughts in the mind. Some scientists love to use the computer analogy: The brain

---

73. John C. Eccles and Daniel N. Robinson, *The Wonder of Being Human: Our Brain and Our Mind* (Boston: New Science Library, 1984), 36.

74. This is the way the biologist J.B.S. Haldane worded it in *Possible Worlds and Other Essays* (Harper and Brothers, 1928), 209. Later C.S. Lewis would follow him.

supposedly works in the same way as a computer operates, since both use a binary code based on "ones" (1) and "zeros" (0); neurons either do (1) or do not (0) fire an electric impulse—in the same way as transistors either do (1) or do not (0) conduct an electric current. This makes it look like the brain "thinks" the way a computer "thinks." I would say there is something wrong here! Whatever is going on in the brain—say, some particular thought—may have a material substrate working with a binary code, but it would not really matter whether this material substrate works with impulses, as in the brain, or with currents, as in a computer, for the simple reason that this material only acts as a physical carrier. A computer may be a relatively accurate simulation of the mind—with a greater or lesser degree of resemblance—but it is not the real thing.

Once we acknowledge that the same thought can be transported by different vehicles—such as pen strokes on paper, currents in computers, impulses in the brain—we realize that a thought is not identical to its carrier. If I were to break my radio, the news report would stop, but this does not mean the news was created by the radio; it was only the news vehicle that broke down. So it seems to me that the brain does not create thoughts but merely transports them. The thoughts somehow "use" the vehicle. The brain is a vehicle of thoughts coming from the mind in the same sense as a book or a CD can be a vehicle of thoughts created by someone's mind. When the late neurosurgeon Wilder Penfield likened the brain to a computer, he did not believe that man *was* a computer, rather, that he *had* a computer—and that this computer has a programmer. He said, "There is a switchboard operator as well as a switchboard."[75]

We tend to say about people with diseases such as Alzheimer's that they have "lost their mind," but what they did lose to a certain extent is their brain, not their mind. Somehow their switchboard is broken; the material carrier for the mind's immaterial thoughts has been damaged. A broken brain is as physical as a broken bone. If these people did not have a mind, they would indeed merely be a broken-down machine that can be discarded at any time. On the contrary, in cases like these, the mind has become an "incarcerated" mind working with a failing brain. So they do not have a "broken mind" but a "broken" brain. To use my previous image

---

75. Wilder Penfield, *The Mystery of the Mind, A Critical Study of Consciousness and the Human Brain* (Princeton, N.J.: Princeton University Press, 1975), xiii.

again, the news is still there but it can no longer come in through a broken radio; there is still news but we cannot clearly hear it. Fortunately, therapy may often help such people to stay in touch with their "self." Or perhaps we should change the roles and say with William Shakespeare in his *King Lear*, "We are not ourselves when nature, being oppressed, commands the mind to suffer with the body."

Fifth, think for a moment how a neuroscientist could ever study the brain. My question would be: Could the brain ever study the brain all by itself? That would be like the magic of a projector projecting itself or a copy machine copying itself. That which studies the brain must be "more" than the brain itself—in the same way as Watson and Crick must have been "more" than the DNA they discovered and studied. To put it in more philosophical terms, the *knowing subject* must be "more" than the *known object*. So I would say that only the mind as a knowing subject is able to study the brain as a known object, because it requires a mind to understand the brain, as it requires a subject to study any object.[76] When science studies the brain, such can only be done thanks to the mind of a subject, the scientist. When studying the human brain as an object of science, a scientist needs the human mind as the subject of science—for without the human mind, with its intellect and rationality, there would be no science at all.

The physical world can never be studied by something purely physical, any more than DNA could ever discover DNA! That is the reason why biology can never fully comprehend the human mind because biology itself depends on the working of the human mind. Biology can deal with the body, including its brain, but not with the mind. The mind can never fully understand itself (neither can the eye view itself). Since animals do not have a mind, they cannot study the brain. Because the mind does not occur on the physical map of the body, it is nowhere in the chain of bodily activities—and yet it is the "soul" of it all and pervades the entire body. We could only objectively study the mind if we were able to stand outside and above the mind, peering down and taking notes, but the thing which is being studied is also the "thing" which is studying it. Sir John Eccles

76. Max Planck said something similar: "Science cannot solve the ultimate mystery of nature. And that is because, in the last analysis, we ourselves are a part of the mystery that we are trying to solve." In: *Quantum Questions*, ed. Ken Wilbur (Boston: New Science Library, 1984), 153.

relates how his teacher, the Nobel Laureate and neurophysiologist Charles Sherrington, told him five days before he died, "For me now, the only reality is the human soul."[77]

Sixth, it is the mind—this mysterious intellectual part of the soul—that allows us to be rational and moral beings. This fact separates us from the animal world. As we discussed earlier, it is not so much the brain that sets us apart as it is the rationality and morality of the mind. One can have more or less brain, but not more or less mind. If the mental were the same as the neural, thoughts could never be right or wrong and true or false, as neural events simply happen, and that is that! In addition to five senses that give us access to a world of material things, we have two more "senses" that give us access to what is immaterial—a world of what is true or false in a rational sense and what is right or wrong in a moral sense. Thanks to the mind, we have a rational sense of true and false as well as a moral sense of right and wrong.

Seventh, by stressing that the knowing subject is "more" than any known object, I want to make clear that all knowledge of objects is based on a subject that apprehends those objects, whereas this subject itself can never be fully captured by making it an object. I-as-a-subject (I-now) can reflect on I-as-an-object (I-past). I can remember my past because I-now is always "more" than and "ahead" of I-past. As a subject, I may investigate I-as-an-object and then realize, for instance, that I-as-an-object made a mistake. In contrast, I-now is never open to investigation because its future possibilities are beyond its current actualities—and therefore, I-now is always a pace ahead of I-past. No matter what I think when I am thinking, it is always I-now who is thinking something, even something like I-past.

We can objectivize anything we want, but we cannot objectivize I-as-a-subject, since that is the very one that has the capacity to objectivize. Put differently, "I-as-a-subject" is not an object like other objects in this world, but it is their very origin; without it, there would be no objects of knowledge comprehensible to us. I can never blame my glands or my animal ancestry for what "I" do wrong, because I-now is always a pace ahead of I-past, including my glands. "I-as-an-object" is my body, which may appear in the iron grip of determinism, but "I-as-a-subject" is my mind,

77. J. Eccles, *Facing Reality: Philosophical Adventures of a Brain Scientist*, (London: Longman; New York: Springer-Verlag, Heidelberg Science Library, 1970), 174.

free to take a next step, a step ahead of any physical determinism, by acting as a new cause with a new effect. I am always more than my genetic code, for I could have an identical twin, with the same genetic code, but I would never be him or her.

Eighth, even the fact that certain mental phenomena are associated with certain neural phenomena does not entail that these mental phenomena were *caused* by neural phenomena. Correlation does not automatically equal causation. The fact that certain regions light up on an fMRI[78] does not explain whether this lit-up state is causing a certain mental state or just reflecting it. It could very well be, as we discussed earlier, that the brain is just a material carrier for the mind's immaterial thoughts. Although something like pain can be induced physically, there is no evidence that experimental stimulation of specific neuronal areas would produce a specific mental state or a specific thought.

When I say that thoughts cannot be physically induced, I am not talking about something like emotions or feelings (even animals have those), because those are physical and biological phenomena that can be physically induced by stimulation of certain brain areas. This is also true of memories stored in the brain—including memories of thoughts once produced by the mind—because memories can be physically stored, similar to the way thoughts can be "stored" on paper. Thoughts, on the other hand, cannot be produced in a physical manner, let alone by electrodes. If the thought of "two times two" would physically produce the thought of "four," we could have skipped much time in school. Brain processes are subject to physical causation, but thoughts are subject to logic and reason.

Ninth, neural activity not only fails to be a sufficient condition for mental activity, but it may not even be a necessary condition. Put differently, mental activity does not always correlate with neural activity; in fact, there may be mental activity even when there is no or hardly any neural activity. We know of situations where the most intense subjective experiences correlate with a dampening—or even cessation—of brain activity. In particular, there can be high mental activity without a corresponding high neural activity. What comes to mind are cases of Near-Death Experiences (NDEs) or Out-of-Body Experiences (OBEs) induced by G-LOC, cortical

78. fMRI is an MRI procedure that measures brain activity by detecting associated changes in blood flow.

deactivation through the use of high-power magnetic fields, mystical experiences induced through hyper-ventilation, and brain damage caused by surgery or strokes.

For all of the above reasons, I consider body and mind two very different aspects of the same human being; you can *tell* them apart but you cannot *set* them apart. The body and its brain belong to the world of objects, whereas mind and soul are part of the world of subjects. These are two different aspects of the same human being—two aspects that we can distinguish but not separate.[79] The fact that we distinguish them does not entail that we can separate them, any more than the idea of a three-dimensional space means that we can separate those three dimensions.[80] And yet they are both real; there is a mental aspect of reality and there is a physical aspect of reality. Yet, as to *how* they interact, we do not really know.

This is not as strange as it may sound. Think of the relationship between mass and gravity; we do not really know how these two interact. Or take the case of electrically charged particles that interact with each other through the mediation of electromagnetic fields; the charged particles affect the fields and the fields affect the particles, but we do not know anything about the "mechanism" behind this interaction. Something similar happens when it comes to body and mind; we know *that* they interact but do not know *how*.

Well, it is this very mind that creates *purposes* in our universe, which is a feature unique to humans. We are undeniably living in a purpose-driven world, driven by purposes in the minds of its inhabitants. As I mentioned earlier, purposes lift us above the rigidity of law-and-order, of cause-and-effect, by creating a new realm of causes, not of a physical but spiritual nature. Are these spiritual causes really something "real"—or are they mere illusions or neural artifacts?

Let us consider the following case. Anyone looking at paintings of the famous Renaissance painter El Greco will notice that most of his figures

79. "How brain and mind interact is probably an insoluble problem," according to Karl Popper and John Eccles in their book *The Self and the Brain* (Berlin: Springer, 1978), Preface.

80. The University of Delaware physicist Stephen M. Barr uses this comparison in a book review in *First Things*, 199 (Jan. 2010), 54-55.

are unnaturally tall and thin.[81] Did El Greco paint them that way "on purpose"? Advocates of scientism would be eager to deny this and then follow the ophthalmologist who proposed in 1913 that El Greco may have had a form of astigmatism that distorted his vision and led to elongated images forming on his retina. I am afraid they are jumping to conclusions too quickly. Why?

Even if El Greco did see the world through a distorting lens, the same distortion would apply to what he saw on his canvas. These two distortions would cancel each other out, and the proportions in pictures would remain realistic.[82] So we must come to the conclusion that El Greco's figures, particularly the holy ones, appear unnaturally thin and tall because that was his intention—he painted them that way on *purpose*! We cannot get rid of purposes and intentions that easily. They are something real and do have consequences, sometimes even of a physical nature.

You might object that purposes are odd entities in a universe of "law-and-order." I cannot blame you. But remember, order is something "odd" too. It cannot be experimentally verified by science and it cannot be proven in other ways. It is something that we, including scientists, must *assume* for the sake of intelligibility, as we found out earlier. Even randomness—which is by some erroneously taken as the opposite of order—cannot be proven. Testing large samples statistically for randomness can never be conclusive. On the one hand, a sequence of events can exhibit an obvious pattern despite actually being produced by a random process; on the other hand, a sequence such as the digits of the number *pi* can pass all statistical tests of randomness without them being independent of each other. Apparently, we cannot prove randomness but must *assume* it in order to understand this universe. In a similar way, we cannot prove the existence of purposes in this universe. It is again something we must *assume* so we can make certain events in this universe intelligible for our minds.

---

81. I borrowed this example from the Nobel Laureate and biologist Sir Peter Medawar, *Advice to a Young Scientist* (New York: Basic Books, 1979), 9.

82. The same argument applies to Vincent Van Gogh's preference for yellow colors. Attributing this to a visual disorder (*xanthopsia*), even if caused by drugs, or to glaucoma, would not make any sense. Even if he did view the world through a yellow filter, he would also view the colors on his canvas that way. Instead, he must have just chosen yellow on purpose!

There is no denying that human beings have the ability to steer the laws of nature in a certain direction, according to certain purposes they have in mind. Where does the power of these purposes come from? The answer is: from the mind. And where does the power of the mind come from? My answer would be: from its Maker. We are rational beings with a mind, because our Maker is a rational Being. We are a reflection of God's mind, made in God's image; as a consequence, we are beings that can make rational (but also irrational) decisions as well as moral (but also immoral) decisions. And decisions, although immaterial, do have consequences, even material ones.

The 19th century biologist Thomas H. Huxley once said about the universe, "The chess-board is the world, the pieces are the phenomena of the universe, the rules of the game are what we call the laws of Nature. The player on the other side is hidden from us."[83] The "player on the other side" is indeed the Great Unknown, but is that "player" really *hidden*? I would say we are able to know something about the Great Unknown due to the fact that our own minds are a reflection of his mind. That is enough for now; more on this later.

At this point, the next question would be: Has the Great Unknown also something to do with the *destiny* of the universe? By "destiny" I do not mean the future of the universe—in the sense of expanding forever (according to the model of an open universe), expanding at a rate slowing down to zero (a flat universe model), or expanding and then collapsing again (a closed universe model).[84] No matter whether the expanding universe does or does not come to an end, this is not the "end" that I am referring to; I am talking about the "end" where physical concepts come to an end, where the physical universe ends and the metaphysical universe begins.

So by "destiny" I mean: Is there a purpose to this universe and is there a purpose in it for me? If there is, then the physical universe could become a meaningful universe—a purpose-driven world. Would such a thing be possible—a meaningful universe? Is there any plan for the universe? How

---

83. T. H. Huxley, *A Liberal Education* (1868).

84. The future of the universe depends on the rate at which the universe expands and the average density of matter in the universe. Knowledge of both of these parameters will tell which of the three models describes the universe we live in, and thus the ultimate fate of our universe.

could the One beyond and behind the universe be of any real help to us in providing for us? Does God really provide, and how can he provide?

These are the kind of questions we are inevitably in for. All religions try in one way or another to lift the veil of the future. How can God provide in a world like ours—closed and running like clockwork? God's providence seems to be a case hard to make. And yet, God's providence is at the core of Judeo-Christian faith!

# 5. Knowing the Great Unknown

Philosophy is capable of showing us that this world would be irrational and incomprehensible without an orderly and lawful Creator—a God of order and reason so to speak. This God is "nature's God"—a phrase, associated with deism.[1] At this point philosophy has reached its boundaries. Religion, on the other hand, is about our *personal* relationship with this Creator God—the God of the Covenant, a God of Love, not only all-present, all-powerful, and all-knowing, but also all-loving. That is where theism[2] comes in. Whereas in deism we are in search of God, in theism God is in search of us.

What do we know about the Great Unknown so far? If God is an Absolute and Perfect Being, he must be all-powerful *(omnipotent),* all-present *(omnipresent),* and all-knowing *(omniscient).* Again, do not take these characteristics as super-traits. God's power is not like our power, raised to the power of a zillion. Instead, God's power is the Source and Origin that all human power depends on and derives from; other powers are not "next" to God but "under" God. We would not have any power if God did not give us some power. And the same holds for God's omniscience; it is not the sum total of all human knowledge, but surpasses it in an infinite way. In that particular sense, God is all-powerful, all-knowing, and all-present. He is infinitely greater than all his works. Let us see what this entails.

## All-Present?

Going through the stages of the previous chapters, I hope the Creator is beginning to take more and more "shape" for you, if I may say so. The

1. Deism holds that reason and observation of the natural world, without the need for organized religion, can determine that the universe is the product of a creator.
2. Theism is a doctrine concerning the nature of a monotheistic God and God's relationship to the universe. It conceives of God as personal, present and active in the governance and organization of the world and the universe.

Great Unknown has become a bit more known. I hope you can now see that one of the things we know about God, the Great Unknown, is that he must be all-present, also called omni-present. We realize that no flashlight, even no telescope or microscope can ever discover God because he is not somewhere present like we are somewhere present. God is nowhere, and yet everywhere at the same time, because God is the origin of all there is, so God envelops all there is. As I said earlier, God is part of everything without being a part of it. Because God is omnipresent, he seems to be "nowhere"; yet he is only seemingly absent, since he is "everywhere" at the same time.

Sometimes they say that God could not be everywhere so he created mothers. I would rather say God *is* everywhere, but mothers are his "hands"; they are the secondary causes without which we could not have been born. They also say about fish that the last thing a fish would ever discover is water; we are in a similar predicament as far as God is concerned. Because God is everywhere, he may seem to be nowhere, but we could not exist outside him or without him.

A classical, biblical way of saying this is that God is the Alpha and the Omega. He is at the beginning but has no beginning, and he is at the end without having an end. A more theological rendition would have it that God is *immanent* and *transcendent* at the same time. God's transcendence—His being infinitely far above all we know—should always be in balance with God's immanence—His being closely and intimately involved with his creatures. Jews and Christians always try to carefully balance transcendence and immanence, so God does not become either too human (similar to deities such as Jupiter) or too distant (like the emotionless watchmaker or absentee landlord of deism). God is part of everything—that is his immanence—but without being a physical part of it—which is his transcendence. God's closeness and God's loftiness are two sides of the same coin. Whoever denies one of these two sides detracts from the Judeo-Christian faith.

Let me explain this important distinction with an example that C.S. Lewis used when he noted that a power outside the universe "could not show itself to us as one of the facts inside the universe."[3] Just as architects are

3. C.S. Lewis, *Mere Christianity* (New York: Harper Collins, 2001), 21.

not part of their buildings, yet are somehow "part" of every part of them, so is God transcendent—that is, not a physical part of what he created—and yet he is immanent—that is, actively involved with each and every part of it. God transcends creation and yet is fully present to it—all-present, that is. We are just God's creatures—not "next" to God but "under" God, not "outside" God but "in" God. "Outside" God, there is nothing.

In other words, God as the Primary Cause, as the Creator of the universe, is not a subsidiary part or element of what he designed; he does not keep a "divine foot" in the door. God can never act like an element or a part of his own creation, similar to the way architects and engineers cannot be a part of the structures they build, even though they are "part" of every part of it. That is also the reason why science can never "bump into" God—not because science denies God (it does not, or at least should not), but because God is outside its framework and therefore cannot be a subordinate part of it. That is where we leave the territory of science and enter the realm of metaphysics and religion.

This view also takes God "off the hook" when it comes to physical disasters in the world. They easily make us question whether God is really in control of this world—that is, whether he is in fact all-powerful and all-present. Indeed, so often the all-mighty God seems to be absent and might-less as to what is going on in the world. In his providence, God is supposed to guide the world towards perfection, but we often get the impression it is actually going downhill instead of uphill. Where is God's providence? Where is his TLC?

It is not easy to answer such questions. Aquinas tried it along these lines. He explains that the way God's providence works is through a hierarchy of causes. God, the universal or Primary Cause of all creation, ordained that the universe would be governed by a series of inferior or secondary causes. In a sense, God is all-present in all secondary causes. One simple example of this is that God made a universe in which small objects would be attracted to larger ones, which we call the force of gravity. Gravity is a secondary cause. By allowing such inferior causes to operate, God made a universe in which he does not have to be the direct cause of every stone falling to the ground. It is through his laws of nature that the all-powerful God is all-present. Therefore, we are able to know what the outcome is of certain contingent events like stones falling. We do not have to wonder about God's will every time a stone falls to the ground, even if it strikes us on the head when it does. God has given us a secondary cause—the force of gravity—which is the direct cause of each stone's earthly plummet.

This way, we live in a world that we can trust because it follows some God-given laws of nature—which is part of God's providence too. Because of his laws, God is all-present. Yet, you may wonder how God's providence allows a stone, following his laws of nature, to hit you of all people on the head. Is God powerless to stop otherwise lawful events from happening? If so, then He would not be all-powerful, and His providence would be rather limited. It looks now as if the all-powerful God has become the victim of his own all-present laws.

Thomas Aquinas, for one, denies this. God could indeed have created a perfect world in which he was in control of everything, but He decided not to. Instead he made a world that is on its way to perfection.[4] Aquinas says that it was in God's wisdom to ordain not to be responsible for all contingent events. And it was also in God's wisdom to allow that there be "defects" in certain secondary causes—such as a falling stone hitting your head or an earthquake destroying your home.

How could God allow such things to happen then? Here is Aquinas' famous answer: If all evil were prevented, much good would be absent from the universe; a lion, for instance, would cease to live, if it could not kill its prey.[5] Or think of something like pain: We could be in real danger if there were no pain to alert us of harm. So God allows "defects" in secondary causes to exist because this contributes to the greater good of the whole, so that the defect in one thing yields to the good of another, or even to the universal good.

Obviously, human beings, as part of nature, are also subject to these contingent events. They feel pain, they get sick, they fall victim to disasters, they get rich and they get poor. Human beings get in the way of all these secondary causes that God has established in His providence. The well-known evolutionary biologist Francisco Ayala rightly places this in a wider context: "As floods and drought were a necessary consequence of the fabric

---

4. The *Catechism of the Catholic Church* speaks of "'a state of journeying' toward an ultimate perfection yet to be attained" (302).

5. Aquinas, *Summa Theologica* I, 22, 2, 2.

of the physical world, predators and parasites, dysfunctions and diseases were a consequence of the evolution of life."[6]

And yet, the Creator does interact with his creation. Isn't this the core of what they call his all-present *providence*? How would this be possible? Is there a way we could try to picture God's providence in a world of "law and order"? Thomas Aquinas may help us here once again. Let me start from an often overlooked statement Aquinas once made: "God is [related] to the universe as the soul [*anima*] is to the body."[7] Elsewhere, he says "the king is in the kingdom what the soul is in the body, and what God is in the world."[8] Notice that Thomas did not say, "God is the soul, and the universe is the body" (for that would amount to pantheism[9]). In his explanation, Thomas uses the analogy of a person—that is, the relationship between mind/soul and body—to portray the relationship between God and world. He does so by way of an imperfect analogy.

What does this image entail? We discovered in the previous chapter that the mind is not found on the physical, scientific map of the body; it may seem to be nowhere in the chain of bodily activities, yet it is the "soul" of it all and pervades the entire body. It is part of everything the body does without being a bodily part. In other words, my mind is the "soul" behind everything I do; it is behind everything my body does. In that sense, we might say it is part of everything I do with my body, and yet it is not a physical part of my body.

Now the question is: Could this relationship between body and mind perhaps help us gain a better understanding of the relationship between God and creation? Let me stress first that this analogy is only an aid to enhance our understanding of God's providence, and therefore inadequate. Perhaps Aquinas never developed this analogy much further because he feared the pantheistic dangers of making God the mind and the universe his body. In pantheism, God would be absolutely and solely immanent in his creation, without remaining fully transcendent to it.

---

6. Francisco J. Ayala, *Darwin's Gift to Science and Religion* (Washington, DC: Joseph Henry Press, 2007), 5.

7. Aquinas, *Sententiae* 17, 1, 1.

8. Aquinas, *De Regno* 1, 12-4.

9. Pantheism is the view that the universe (or nature) and God (or divinity) are identical.

Undeniably, the use of analogies is always limited, and therefore a potential source of dangerous implications. One obvious limitation of Thomas' analogy is that the mind does not create the body, whereas God is the Creator of Heaven and Earth. Another one is that the body does not depend on the mind in the way the world depends on God. So we cannot let this analogy mislead us. Nevertheless, in my view, its advantages seem to outweigh its disadvantages. Therefore let us take a look at it more closely so as to better understand how God might interact with his creation.

The analogy I am using here suggests that God somehow works in the world as the mind and soul work in the body.[10] If so, God's intervention in nature and evolution is not a physical intervention in the way science understands it, but one similar to the way the mind interacts with the body—being part of everything without being a physical part of it. On the one hand, this analogy prevents us from thinking that God's activity is a physical, inner-worldly factor interfering in the physical process of the universe, thus restraining us from degrading God to a secondary, natural cause in the midst of other secondary, natural causes. On the other hand, God can still be involved with what is going on in the world—but without becoming a physical part of it. In that specific sense, God is the soul that pervades all of creation and is "part" of everything in creation.

God may seem to be nowhere in the chain of worldly events, yet he is the "soul" of it all and pervades the entire universe. Since God is the source and soul of everything, he is part of each and every cosmic event that is taking place, but without becoming a cosmic part. God is not one of the players on the world scene, but he definitely is the Author and Director of this "cosmic play." Species appeared on the evolutionary scene because God wrote the "script" that way; he is the direct Cause of every detail of the universe. Although he is not a physical element in human history, he is certainly its Framework—the Lord of history, keeping everything in the world (including its secondary causes) in concert. As the mind and soul are "all-present" to the body, so is God all-present in his creation. As the mind has purposes for the body, so has God purposes for the universe.

The analogy of God as the author of a play, may be helpful here, but it

10. We try to understand here how God provides through the way we steer and design things ourselves. But it is actually reversed: We are only able to steer things because we were made in God's image.

has one important drawback: A play does not necessarily require a playwright. Actors of the play can speak and act on their own without any playwright or script. In other words, a playwright is not necessary for a play—so God would not be necessary for a cosmic play. To avoid this consequence, we could compare God to the author of a novel instead. A novel has no actors, only the story itself. Although one can read a novel without giving any thought to its author, there could not be a novel without an author. That seems to be a more accurate image for the relationship between God and creation. Yet, as with all analogies, this one has another limitation—it makes human freedom an illusion. Hence, I prefer the image of God as the Author of a Cosmic Play.

Given either image, it should not surprise us that God can be found then in everything—according to the famous adage of Ignatius of Loyola. Everything in this world speaks of its Author. God can be found in the birth of a child given to us, in the healing we received through medication, in the successes we achieved in our lives, in the technological tools we successfully construed, as well as in the discoveries science made. God is not in competition with any of these, for everything we experience is "for the greater glory of God," as Ignatius used to say. Just as we can use the laws of nature for our own intentions and purposes, so may God very well work through natural causes and human intentions to achieve his divine purposes and plans for salvation. God works through all of these like our minds work in our bodies, in a way that necessarily eludes science—that is, mysteriously, yet powerfully. God's presence envelops our purpose-driven universe. Let us call it TLC.

## All-Powerful?

Not only is God all-present, he is also all-powerful—but again, as long as we do not misconstrue this term. God's power does not exceed other powers but it *transcends* other powers. It is not a worldly power raised to the zillionth power, but it is of another dimension; it is an other-worldly power—all-powerful, al-mighty, omni-potent. Again, be aware that the Infinite Primary Cause is in no way comparable to finite secondary causes.

Many people have misinterpreted God's omnipotence, though. Probably the best known example is the French philosopher and atheist

Jean-Paul Sartre.[11] He took it that an almighty God does not leave any room for free human beings, whereas free human beings do not leave room for an almighty God. In Sartre's eyes, God and Man seem to be in a power battle. He opted in favor of human freedom over divine omnipotence and thus became an atheist (until just before he died). Earlier we had to defend human freedom against the doctrine of complete, hard-core, ontological determinism; now Sartre feels the need to defend human freedom against what he considers the doctrine of theism. Does Sartre have a point here?

I would say he does not—for at least three reasons. The first reason is that Sartre's dilemma puts God and Man on the same level, whereas they are certainly on very different levels. God is not one of the persons among other persons, just as he is not a cause among other causes. We are not "next" to God but "under" God. Therefore, submitting ourselves to God, the Maker of Heaven and Earth, is not like submitting ourselves to a dictator, who is just another person in our midst. On the contrary, the more we become like God, the more we become like ourselves, since we were made in his image. We tend to think that our freedom is compromised by God's will, so we take it as a threat to our freedom against which we should rebel. We think we are only free when we are free of God. Instead, following God's will would be an opportunity for us to become more like our deepest self.

My second reason for rejecting Sartre's dilemma is that God's omnipotence does not take my freedom away. God's power is not a blind brute force like the forces we are familiar with, but a loving, intentional power far beyond our comprehension. As a matter of fact, God decided in his omnipotence to make us humans "participants" and "co-workers" in his creation, in accordance with his image. He chose in his own freedom to endow us also with freedom, as a reflection of his own freedom. Consequently, creation did not spring forth complete from the hands of the Creator. God also gave us the dignity of acting on our own, and thus of cooperating in the accomplishment of his plan, by enabling us to be intelligent and free causes of our own in order to complete the work of creation.

11. Jean-Paul Sartre asserts in *Being and Nothingness* that man is a creature haunted by a vision of "completion," called by Sartre the *ens causa sui*, which religions identify as God.

My third reason is that "human freedom" would remain hanging in the air if there were no God. We have nothing to base it on, other than God. It certainly cannot be based on genetics. Theoretically, there could indeed be a gene that allows us to make choices, but if this gene or additional genes would also determine the outcome of these choices, then we cannot really make free choices and have basically lost the free will we thought we had. If we want to claim human freedom, then we need someone from whom this freedom derives—a Creator God. So there is no dilemma of "God *or* freedom," but our only choice is their unison—"God *and* freedom."

Apparently, there can definitely be human freedom under an all-powerful God. Dictators may take human freedom away, but God made us in his image and thus he created us, not as marionettes, but as beings endowed with freedom as well. Since God made us after his own image, we cannot just be marionettes or automata. God gave us freedom in his selfless love. Because we are rational beings, made in God's image, we have been created with free will and are master over our acts—the right to choose is ours. God lets the actors on the world stage be free actors, who may not act the way the Author of the play would like them to act. Isn't it striking how he allowed our weak human will to withstand an almighty, all-powerful, sovereign God? God could perhaps have chosen to eliminate the possibility of evil and evil-doing, but then God would have also taken away the possibility of good and doing-good.

It is clear that the possibility of *evil* is also a consequence of this human freedom. How could God ever give us freedom without accepting its consequences up to the point of us freely choosing the wrong outcome, away from God and in opposition to his divine plan? If our world were fully preordained by God, our brains would automatically refuse to enforce what our free minds would like to do. That would be a travesty of human freedom: "a toy world which only moves when He pulls the strings," in the words of C.S. Lewis.[12]

Fortunately—and sometimes unfortunately—God takes his creatures seriously, so they have also the capacity of making the wrong decisions. And that they did and keep doing, including you and me. So, humans do not have to be powerless if there is an almighty God. Sartre was wrong on

12. C. S. Lewis, *Mere Christianity* (San Francisco: Harper, 2001), chapter 3.

that issue and created a false dilemma: No longer do we have to question God's omnipotence, or even his existence, if there is human freedom. The fact that we do have freedom actually points to a Creator after whose image we were made: How could there be human freedom if there were no God who has freely created us after his image?

This leads us to the following question: If God is almighty and all-powerful, does this mean he can do whatever he wants? Apparently he cannot, for he has given part of his power away by giving freedom to human beings. But what about a different kind of power, you might ask: Does he have the power to create a square circle, for example, or can he make a stone so heavy that he himself cannot lift it? Is that not contradictory, or what?

When Aquinas dealt with this question, he was very definite in his answer: When something is against reason, God cannot create it.[13] What made Aquinas so sure on this issue? The answer is that God *is* reason and freedom, so he cannot act against his own nature by doing what is contradictory. God is absolutely free, but his freedom is not arbitrary, so he cannot go against what is true and right. How do we know this? Well, our own power of reason is rooted in creation and thus participates in God's power of reason. If our reason tells us that something like a square circle is impossible, then we see this with our very intellects which were created by God and are participating in God's reason.

So, being all-powerful does not mean being able to do what is logically contradictory. Aquinas gives many examples: God cannot create square circles; God cannot make someone blind and not-blind at the same time; God cannot declare true what is false; God cannot undo something that happened in the past, and the list goes on and on.[14] The same applies to human freedom. Freedom is always a two-way street, leading us *either* toward God *or* away from God. So when God gave us freedom, he could not give us freedom without the potentiality for sin and doing evil—that would be logically contradictory. Therefore, evil can be, and must be, a very real consequence of human freedom.

---

13. Aquinas, *De Aeternitate Mundi* I, 4.

14. Aquinas, *Summa Contra Gentiles* II, 25.

The word is out—evil! People who have found God in their lives usually have at least one more hurdle to take: the problem of evil. Evil has been one of the strongest obstacles for Judeo-Christian faith. To put it in a nutshell: How does an all-powerful God get away with the evil in the world? If God is not able to take evil away, God cannot be all-powerful. And if God is really all-powerful, there should not be any evil.[15] Or should there?

As far as *moral* evil is concerned—that is, evil that humans cause themselves—the answer can be found in the previous paragraphs: Moral evil is a consequence of human freedom. But what about *physical* evil—evil not caused by human beings, evil that seems to come with nature, such as natural death, famine, diseases, earthquakes, tsunamis, and other catastrophes? Several answers have been given to this haunting question.

The first answer is that, if there were no physical evil, such a fact would diminish the good of the universe. To explain this answer, Aquinas again uses the example of the lion that could not live without killing its prey.[16] Whatever may be evil for the individual, the prey, is good for the larger picture, the universe. Consequently, if there were no evil, such a fact would diminish the good of the universe. Environmentalists are very aware of this fact; even 'dangerous' animals such as poisonous spiders and snakes play an essential role in their ecosystem; taking them out would disrupt the system. Another example would be physical *pain*, which is a valuable warning sign telling me that my body is in danger (lepers can certainly testify this). So we need not look at it as evil. However, such an answer would still leave physical *suffering* such as leprosy unexplained. Would such suffering be good for the larger picture as well? That seems quite a stretch, doesn't it?

A second answer would run like this. Nature is bound to follow its God-given laws of nature; well, the laws of nature must operate as they do, if intelligent and free agents are to exist. Therefore, we must distinguish between what God *wills* and what he *allows*. Aquinas says that God "neither wills evils to be nor wills evils not to be; he wills to allow them to happen."[17] God does not will earthquakes, but he allows them when they

15. David Hume had already worded it this way: "Is God willing to prevent evil, but not able? Then he is impotent. Is he able, but not willing? Then he is malevolent. Is he both able and willing? Whence then evil?" in *Dialogues Concerning Natural Religion*, 244.

16. Aquinas, *Summa Theologica* I, 22, 2, 2.

17. Aquinas, *Summa Theologica* I, 19, 9, 3.

are a consequence of the laws of nature—in the same way as God does not will wars but allows them when humans use their freedom to start them. In other words, there is God's positive or *providential* will and then there is God's *permissive* will; therefore, not everything that happens in this universe is directly willed by God's providence. To say that God "allows" or "permits" evil does not mean that he sanctions it in the sense that he approves of it, or even wants it. That is why the Lord's Prayer, the Our Father, says, "Thy will be done on earth as it is in Heaven." Apparently, that is not always the case. What God wills, with and for us, must become the measure of our willing and being.

A third answer would tell us that God's creation is in a state of journeying toward an ultimate perfection yet to be attained. God's creation is not perfect yet, but it is on its way, even through the process of evolution—and we, human beings, have been made participants in his creation; we are his "co-workers" in bringing his creation to perfection. God *wills* perfection but *allows* imperfection on the journey to perfection. So physical evil is part of the imperfection we are still surrounded by—and so is moral evil. Moreover, when speaking about perfection, we should always realize that we can only have a sense or idea of perfection if we look "down" from a heavenly perspective where perfection is located in God's Mind. This is no longer a bird's-eye view but a God's-eye view.

A fourth answer turns the world completely upside-down. When speaking of physical evil, we find ourselves already in a mental, spiritual, even moral context. When speaking of "evil," we are asserting somehow that evil "should" not exist. We are in fact evaluating physical suffering as wrong or bad—something no animal would be able to do. A prey does not consider the predator "evil"—perhaps painful, literally, but not evil. When giving birth, animals may experience physical pain but not suffering in the sense of something "bad" or "evil." Humans, on the other hand, do! As a consequence, physical events only become evil when we assess them as bad and label them as evil. That label is ours.

So the "thorns and thistles" may have always been there, but since the Fall in Paradise, they were felt not only as painful but also as distressing, as something "evil." Aquinas makes a very astute remark in this context: "Some say that the animals, which are wild now and kill other animals, were not that way [in paradise ...]. But this is entirely unreasonable. The nature of animals was not changed by the sin of man."[18] Aquinas is right; after the Fall the *world* did not change, but *we* did. Hence, even the cause of physical evil is ultimately sin. Without sin, physical evils would not rankle or embitter us. We are oppressed not so much by suffering or the wretched conditions of our existence as by sin. That is where the priority lies in our need for redemption.

I must say there is something to this last answer, like it or not! Only humans take diseases and catastrophes as something that should not be, as something that seems to be acting against them personally. Only humans can get depressed. Animals may "dislike" these things, but they do not question them in terms of "Why *me*?" They do not have a "me," and since animals do not know about good and bad, they cannot ask why bad things happen to good animals. Only humans know of God, so they ask the question "Is something wrong between God and me?" or "Why do bad things happen to good people?"[19]

Let us not forget that, without God, we could not even speak in terms of evil. Thanks to God, we know what the world "should" be like if we look at it from a "God's-eye view." We know of "evil" because we have an idea of "good" and of what things should be like, if everything were "good"—the way God intended them to be. Interestingly enough, Jesus of Nazareth, the human face of God on earth, was never viewed as the one to *cause* anyone to be blind, to be lame, to become a leper, but rather as the one to *cure* people from blindness and leprosy, because such afflictions show the absence of something good that should have been when seen from a Heavenly perspective. Evil—and hence suffering—has everything to do with sin. When a paralytic was brought to him, Jesus said, "your sins are forgiven" (Mk 2:5). The paralytic wanted to be able to walk, not to be delivered from his sins. But Jesus saw the man's real need. Forgiveness of sins is the foundation of all true healing.

This takes me to the fifth and last answer—actually not an answer in the technical sense. It does not come from words but from a person called Jesus of Nazareth. It is an answer that comes in fact from the Man of Golgotha: God is love—and love wants to share to the very end, with

18. *Summa Theologica* I, 96, 1, 2.

19. This is a reference to Rabbi Harold Kushner's book entitled *When Bad Things Happen to Good People* (New York: Random House, 1978).

all its consequences. God's love wants to share everything with us, even our sufferings. God cannot suffer—he is *impassibilis*— but God can "suffer with"—he is not *incompassibilis*.[20] So Jesus came, not to abolish, but to sanctify suffering with his presence. Jesus did not save us *from* the cross, but instead he saved us *by* the Cross. Even in suffering—or particularly in suffering—we can find the Glory of God, for Jesus is the human face of God—and a human face comes with tears. Seen from this angle, even suffering can at times become a blessing.

Am I glorifying suffering? I do not think so. Yet, I realize we live in a world that runs away from suffering; our bathroom cabinets are filled with painkillers. Since the time of our youth, we have been conditioned to view suffering as an impediment to happiness. This worldview, which is so embedded in our culture, tells us that the less we suffer, the happier we will be. Yet, we could be missing out on another dimension of suffering, for suffering has this mysterious potential of redeeming us, transforming us, transfiguring us. You might think the less we suffer, the closer to God we will be—but it might actually be the opposite. Suffering can be very therapeutic.

In spite of all of this, most of us do have the opinion that a perfect world from the very beginning would be better than an imperfect world on its journey to perfection. We tend to assume that God's providence should have created an ideal world, a hedonistic paradise, a place in which comfort and convenience are maximized, a world in which everyone has an electrode implanted to cause intense euphoria and ecstasy with a simple push of the button. But I ask you, do we really admire those who appear to have a life of ease? What we do admire instead are lives of courage and sacrifice; we have a high regard for people who overcome hardship, deprivation, or weakness so as to achieve some notable success; people who stand against some great evil, or who relinquish their own happiness to alleviate the suffering of others—in short, people who take up their crosses. A life without challenge is a life without interest. Apparently, the maximization of creaturely pleasure does not seem to be a top priority in most lives—let alone in God's omnipotence and providence for us.

---

20. Bernard of Clairvaux, Sermon 26, no. 5.

## All-Knowing?

Calling God all-knowing usually evokes responses very similar to the ones connected with his omnipotence. If God is really all-knowing, then his knowledge must include the very future that we are trying to shape ourselves. If that is the case, then one of the two has to go: either an all-knowing God or a free human being!

No wonder this issue has caused much turmoil in the history of Christianity. This time it is not atheists who are forcing us to make a choice between God and Man, but Christians. If God is all-knowing, he must have foreseen our future. So some Christians say he must have foreseen that some will end up in Heaven and others in Hell. If God knows everything, he must also know who will be saved and who will be damned. If God knows everything, then human freedom must be under attack again. My response would be: Or perhaps not! This issue is usually referred to as *predestination*. We will get back to it later on in this chapter.

First we need to find out what makes us think that God is all-knowing or omni-scient. Following the early Christian writer Boethius, Aquinas held that God's position with respect to time is such that, unlike us, he does not have to wait for the future to unfold in order to know its contents. God is in no way a temporal being, but he is rather the creator of time, with complete and equal access to all of its contents.[21] If he were in time, even God could not know what has not yet happened, since that would be contradictory. But if God exists entirely outside of time—in a kind of eternal present to which all that occurs in time is equally accessible—he would indeed be able to comprehend all of history, the past and the present as well as the future, just as though they were now occurring. According to this view, an all-knowing God would have fore-knowledge of future events, decisions, and actions because his foreknowledge is situated outside time, external to time. It comprises all that ever was and ever will be.

The question remains, though, how God can know which decisions and actions his creatures will engage in, without losing the idea that those decisions and actions are done freely. How can God *foresee* our future

21. Time is a product of creation, for God brought into being both the creature and time together. That is why Albert Einstein could show us that time and space are part of the physical world, just as much as matter and energy.

response to his call for salvation, if we want to remain free human beings who can make free decisions in life, including a free response to God's call for salvation? How can he know "ahead of time" what we will end up doing if we are free to make our own choices in life?

The answer is actually rather straightforward. Perhaps a simple analogy may help us understand that this is not a contradiction. If you were watching a video of certain events of your past life, you may get the impression that these actions were in fact predetermined, and yet you know that they were freely decided upon when they were taking place. Well, God in his eternity is like someone watching the "video" of what is taking place on Earth without taking any freedom away from the "actors on the stage." Again, this is just an analogy, and therefore inadequate.

Aquinas would explain the apparent conflict between God's activity as creator and ours as free creatures by again using his powerful distinction between Primary Cause and secondary causes. We are prone to think of God's activities as some kind of secondary cause, situated at the same level as our own free decisions. We think, for instance, that God creatively wills that I decide to do something, and his willing then causes me to decide—as if his will were a secondary cause. In this scenario, God's creative fiat would be an event independent of my decision, which would indeed rob me of my autonomy. I would not be a free agent anymore but a puppet manipulated by God.

In contrast, God as a Primary Cause works in such a way that we are not acted upon as if God were a secondary cause, but instead exercise our own free will as a secondary cause of our own decisions and actions. If this view is right, then our human freedom need not be in conflict with an all-powerful and all-knowing God at all. God is the complete and Primary Cause of a free act, whereas the human agent is its complete secondary cause. Being all-powerful as a Primary Cause, God exercises full control in all that we do, notwithstanding the fact that our deeds are fully voluntary; so we still have every reason to expect that all that takes place in the world will reflect the providence of an all-loving Father. God's willing that I decide as I do does not make my decision God's. God's willing does not take away from me the operations of my will, or the actions founded upon them; they remain my own.

In the end, it really matters what kind of person you and I have been.

The secondary causes of my decisions really matter, as much so as all other secondary causes do. They all work under the guidance of the Primary Cause. In the end, human freedom still stands tall: We can only be rational and moral beings because we were created with a free will—the right to choose is always ours.

I think we may conclude now that there is plenty of space for human freedom in this universe—in spite of contrary claims made by ontological determinism and an incorrect conception of God's Sovereignty. Yet, it seems to me that we have also reached a point here where human freedom has reached it limits. Often we bang our heads against a wall of stubborn events, events beyond our own power—events such as failures, disasters, afflictions, you name it. It is at those moments, when we feel down, or worried, or even crushed by a life-threatening blow, that religious people may come up with this beautiful encouraging response: Do not worry, God will provide.

What do they mean when they say this? How do they know this is true? Probably, it is their Church that told them. But how then does the Church know that this is true? And if it is true, we wonder why so often we do *not* see God providing. We keep waiting and waiting, or we notice that God seems to provide for other people—but not for us. One question leads to another question. Could it be that some people are his favorite ones? Were some people perhaps chosen and others rejected from the start? Believers have been struggling with such questions for centuries. Sometimes they came up with shocking answers. You might think of certain Christians, mainly of the Calvinist tradition, who seem to tell us that some are predestined—actually predetermined—to Heaven and others to Hell. That does not look like a providing God, does it? There are many, many questions, perhaps too heavy and serious for a book like this. But let us give it a try.

# 6. All-Loving Providence

Once we accept that God *can* provide, the next question is whether he *does* provide, and how. We have reached a stage now where deism cannot help us any further; it just leaves us alone in the cold of interstellar space. Deism would admit that the world was made by God, but as by a watch-maker who makes a watch, and then abandons it to itself—a "hands-off" approach, so to speak, of an absent landlord.

I hope I have cleared the path enough now for a more personal relationship with God. Whereas in deism we are in search of God, in theism God is in search of us. It is time to replace "the God of philosophers," as Blaise Pascal called him, with "the God of Abraham, Isaac, and Jacob."[1] Well, it is this very Abraham who takes his son Isaac to a mountain so as to sacrifice him to the Lord, and then is asked by his son, "Where is the lamb for the burnt offering?" Abraham answers him, "God will provide for himself the lamb for the burnt offering, my son."[2] And then, when the angel of the Lord stops him from sacrificing his son, Abraham finds a ram in the thicket. So he calls this place "The Lord will provide"—which means, God will take care of it.

Many people have been inspired by this story and have learned that there is always "a ram in the thicket," when you least expect it. As the apostle Paul puts it, "my God will supply all your needs."[3] If so, we are indeed living in a purpose-driven universe, which is also driven by purposes in the mind of its Creator. He is not an absentee landlord but an all-present landlord. So my question is: Is he really?

1. Blaise Pascal, *Oeuvres complètes* (Paris: Seuil, 1960), 618.
2. Book of Genesis 22:8.
3. Letter to the Philippians 4:19.

## A Universe with a Destiny?

In contrast to deism, theism tells us that the Creator of this world is also, or first of all, an *all-loving* God who created the world out of love. He has a plan for His creation, a plan for all he created, including you and me. When the apostle Paul was in Athens and saw an altar inscribed "To an Unknown God," he invited the Athenians to take the step from deism to theism, from a God of reason to a God of faith: "What therefore you worship as unknown, this I proclaim to you."[4] Paul wanted to make the Great Unknown known as an all-loving God.

What did Paul know about God? His message goes along these lines: Because God's creation is in a state of journeying toward an ultimate perfection yet to be attained, God guides his creation toward this ultimate perfection with divine and loving *providence*. Even evolution unfolded exactly as known, planned, and willed by God from all eternity. God's creation is not perfect yet, but it is on its way—and we are invited to be God's co-workers, participants in his creation. As co-workers we can bank on God's providence, so the universe can reach its destiny.

But do not take this assurance as a blank check. Providence is not a divine insurance policy. A religion that promises immediate answers to prayer and instant awards for good behavior is merely a commercialized version of religion. God is not a cosmic vending machine! God does provide, but there is also a personal cost on our end. Do not expect God to bail us out of situations of our own making. Yet, in prayer, we always get what we need—not what we *think* we need but what we *do* need. God supplies all our "needs," but not all our "wants."

When the apostle Paul says that God will supply all our needs, he is basically saying the following: Do all you can and let providence take care of all the rest. Apparently, there are always two sides to the Good News story: We have to do *our* part in order for God to do *his* part. There is obviously a strong interaction here. Since we were given freedom, we have to use the gift that was given to us. Free creatures are of greater value than puppet creatures, because their greater likeness to God would be an improvement to creation. When we do our part, we leave the rest up to God's providence. We may not know what the future holds, but we do know that God

4. Acts of the Apostles 17:23.

holds the future. The future of the universe is in the hands of its Maker. Everything in this universe unfolds in accordance with God's plan. God orders all things, whether the falling of a sparrow or the hairs of our heads, which are numbered.

As I said already, even when we do our part, it may not be fully matched by God. We all know we cannot just sit home and expect God to pay all our bills. For we too have to do our "homework." And yet, do not expect that everything will unfold the way you had planned it. Make your plans, but keep in mind God may have something else in store for you. As the Bible puts it, "The mind of man plans his way, but the Lord directs his steps."[5] There is also this well-known classic line from Thomas à Kempis, "Man proposes, but God disposes,"[6] which means human beings may make any plans they want, but it is God who decides their outcome. Even Woody Allen knew that: "If you want to make God laugh, tell Him your plans." This may look sort of de-motivating, but the good part of it is that, even when we do bad things, God can turn them into something good according to his greater plans. The Apostle Paul says it even better: "we know that God causes all things to work together for good."[7] He is definitely referring here to *all* things, even if they include bad things.

At times we do see God's guidance and providence in our lives—for instance, when we met our spouse for the first time, or when we gave birth to a beautiful baby, or when we got that rewarding job, or when we could start a new enterprise that was successful—but at other occasions, we just feel surrounded by mere fortuitous events, mere coincidences. The question remains, though, whether those seemingly senseless events are really coincidences.

We tend to adorn the word chance with a capital C—taken as "blind fate" and "good or bad luck"—which turns life into a mere "hit or miss" process. But whatever appears to us as mere "coincidence" may in fact very well be a directed, indeed providential process when seen from a religious point of view. The concept of randomness in science is about the relationship between secondary causes, but it has nothing to do with how secondary

---

5. Book of Proverbs 16:9.
6. Thomas à Kempis, *The Imitation of Christ*, Book I, chapter 19.
7. Letter to the Romans 8:28.

causes are related to the Primary Cause. As to how things in the universe are related to the Maker of the universe is a completely different story.

Padre Pio was apt to say in various ways, "God arranges the coincidences." Once he asked a man who claimed such-and-such event had happened by chance: "And who, do you suppose, arranged the chances?" Science has no answer to this question—not even the answer "nobody did." The physicist Stephen Barr is right when he said, "No measurement, observation, or mathematical analysis can test whether or not God planned a development like a genetic mutation. What apparatus would one employ? Being 'unplanned by God' is simply not a concept that fits within empirical science. Being 'statistically random,' on the other hand, is, because it can be tested for."[8]

Who arranged the chances? I would say the answer to that question is… God. There is a divine dimension to our universe; what unites and defines us as human beings is not "being of the species *Homo sapiens*" but being "under God." Everything that befalls us is a gift from God—and that is where providence significantly surpasses serendipity. In religion, there is no place for good or bad "luck." As a matter of fact, Aquinas once said, "Whoever believes that everything is a matter of chance, does not believe that God exists."[9] Just as we on our own can manipulate the laws of nature to achieve our own purposes, so does God work through natural causes and human intentions in order to achieve *his* purposes and salvation plans, because he is "part" of all that happens in this universe, without being a physical part of it. It is God's providence that allows us to see events in their inter-connectedness.

Believers of the Judeo-Christian tradition know very well that God does work this way—constantly, faithfully, and lovingly. Elsewhere, Aquinas says, "the causality of God, who is the first agent, extends to all being. […] It necessarily follows that all things, inasmuch as they participate in existence, must likewise be subject to divine providence."[10] The Bible shows us that God's involvement with his creation is really miraculous. As the Book of Proverbs says, "The lot is cast into the lap, but its every decision is from

8. Stephen M. Barr, "Chance, by Design," *First Things*, 2012, 12, 26.

9. Aquinas, *De Symbolo Apostolorum* IV, 33.

10. Aquinas, *Summa Theologica* I, 22, 2.

the Lord."[11] Anything that seems to be random from a human point of view may very well be included in God's eternal plan. It is God himself who rolls the dice, so to speak, for "its every decision is from the Lord." Therefore, do not worry, for God provides and will take care of everything. Mahalia Jackson expressed this so well in her song "He's Got the Whole World in His Hands." God is the direct cause of every detail of the world—whether the falling of a sparrow or the hairs of your head, which are numbered.

There is one more point I should make in this connection. Instead of saying that we should do all *we* can, and then let providence take care of all the rest, we could also reverse the order: Providence gives us chances and then we should do our best to grab those chances. This reversal takes us into the classic discussion of *grace* versus *works*. Some have made this issue into an either-or dilemma by putting human freedom against God's sovereignty again—so that one of them is supposed to give. Whereas the heretic Pelagius held that we alone are responsible for our salvation (through our works), the reformer John Calvin in particular argued the opposite—that God alone is responsible for our salvation (through grace).

I do not think this is a true dilemma. It is clear that without grace we cannot do anything, since all secondary causes are fully dependent on the Primary Cause; without creation there would not even be any creatures. Good *works*, on our side, are always a fruit of *grace*, coming from God's side first. Jesus emphasizes this with the imagery of the vine. "I am the vine, you are the branches [...] apart from me you can do nothing."[12] God's part comes first, of course. However, on our side, we are free to cooperate with God's grace or not. His grace is captivating but not irresistible.

In saying this, I do not minimize the need for God's grace, but I do recognize that God's plan includes the possibility that we could throw away the gift we have been given. Grace, which literally means "gift," is God's favor to us. In other words, it is not something received in return for anything given from our side—it is free and unmerited. Everything is grace, and without it, we would fail. Without God we could not do anything—we could not even exist! We could not love if we were not first loved by God, but it also remains true that we are called to love in return.

---

11. Book of Proverbs 16:33.

12. John 15:5.

So when people say that our salvation depends on what *we* do, they should realize that salvation can only come from God's grace, not from our doings. Hence, we do not *earn* salvation, for it is a gift based on grace, but we do have the choice to accept or reject that gift through our works. So we need to do our part, our works, to let God's grace work. What *we* do does not take anything away from what *God* does in his sovereignty. If God had not given us the gift of free will, we could not be rational and moral beings who can make rational and moral decisions. As you can identify a tree by its fruit, so you can identify people by their actions, says the Gospel.[13]

However, trees need to be cultivated to bear good fruits, so they need hard work. If you want to enjoy a beautiful garden or orchard or vineyard, you cannot just sit down, waiting for the fruits to come out good. Good fruits require constant care on our part, but the power to grow comes from God. To put it differently, God has no hands but ours; and we have no power but his. However, never put your work in opposition to God's grace. They are not competitors but they grow in union. Ignatius of Loyola used to put this in the following terms: Work as if everything depends on you, but trust as if everything depends on God. The problem with the idea of providence is that when you push it too far, you will start wondering if free will really exists. But when you ignore it, God starts looking like an absentee landlord.

Then there is another important question that needs to be addressed: Are we really free to reject God's grace? The answer is a definite yes. God gave us this freedom, as we discussed extensively in the previous chapter. We are indeed free as to whether we accept grace and do good, or whether we reject it and do evil. If Adam and Eve could fall from grace, certainly all of us can fall from grace as well. How is this possible, you might ask, given God's sovereignty. Well, God provides sufficient grace for everyone to be saved, but whether it is "efficient" and put into effect (efficacious) depends on *us*. Aquinas would say "grace changes the will without forcing it."[14] God offers salvation to all, but this may not produce salvation for all.

But no matter how we understand this, God's grace is not irresistible like some Christians believe. Therefore, God does not choose the saved

13. Matthew 7:20.

14. Aquinas, *De Veritatis*, 22:9.

ones, but he does save those who choose to accept his grace. If grace were irresistible, God would be forcing people against their will to come to him. That is not God-like, and certainly not biblical. Yet, it may sound strange or even offensive to some that the sovereign God of Christianity made himself dependent on Mary's fiat so his Son could be born in her womb. Nevertheless, that is the way God works. The only way God can redeem man, who was created free, is by means of a free "yes" of a human being. God's power is tied to the unenforceable "yes" of a human being—Mary—and human beings like you and me.

## A Universe of Good and Bad People?

Yet there remains this enigma of human suffering. Why is suffering such an immense problem for the Judeo-Christian religion? The reason for this is pretty simple when you look at it from a religious perspective: How can there be suffering if God always provides? In addition, it is a problem when seen from a logical perspective: If God is so good, why is his world so bad? If an all-good, all-wise, all-loving, all-just, and all-powerful God is running the show, why does he seem to be doing such a miserable job of it? All religions, Christianity in particular, are obligated to address these crucial questions.

First of all we have the problem that afflictions are distributed so *unpredictably*: They strike the just as well as the unjust, believers as well as unbelievers, the good and the bad alike. There is no pattern! We all seem to have the same chances to be stricken by evil and suffering; no one is exempt; even one's religion does not seem to make a difference. Actually nothing seems to make a difference. Suffering could well be called the truest democratic experience of them all.

In addition, we have the problem that afflictions are distributed so *unequally*: Some people have to stomach so much more than others; some receive one blow after another, whereas others are apportioned poorly. At times, you meet people who remain erect in the hurricane of misery; then again, you come across people who lament endlessly about trifles. We were created equal, but surely not as far as misery is concerned.

Anyone eager to build a system explaining all of this will eventually be buried under a collapsing house of cards. Nothing seems to fit, nothing seems to make sense. And yet, religion is the only place where we could search for an answer. Whereas suffering may be as painful to Humanists, Marxists, and Buddhists as it is to Jews and Christians, only the latter are haunted with this piercing question, "Why does God abandon me?" Believing in a God of love, believing in a good creation, believing in the providence of an all-loving God, causes the pain of suffering to penetrate to a deeper level—to the level of "Is something wrong between God and me?" In response, the Gospel of John tells us, "God sent the Son into the world, not to condemn the world, but that the world might be saved through him."[15] As I said before, Jesus, the human face of God on earth, was never seen to *cause* anyone to be blind, to be lame, to become a leper, but he would rather *cure* people from blindness and leprosy, because such afflictions show the absence of something good that should have been.

So is there something wrong between God and me then when afflictions strike me? To answer this question, let us first rephrase the question: If I ask myself why evil strikes *me*, I could ask myself as well why evil would *not* strike me. Realizing suffering is everywhere may help us to de-center from our own suffering. Or consider the example of an infection. Each time we get an infection, we tend to be shocked, but we should rather be surprised that we usually do *not* get an infection. The real wonder comes from our beautifully designed immunity system—another wonderful piece of creation, but perhaps not perfect yet. But as God's co-workers, we may be able to find some medical cures.

The same is true of the tantalizing question as to why bad things happen to good people. We could as well ask the reversed question—why do good things happen to bad people. Who is to say we are good people anyway? Are we not all sinners? Jesus would say: Let those who are without sin throw the first stone. To paraphrase G.K. Chesterton, as far as sin is concerned, there are two kinds of people: not, as you might think, those who sin and those who do not sin, but those who know they are sinners and those who do not know they are sinners.

In other words, not only is there much good in the worst of us, but also much bad in the best of us. Looking at things this way gives us a completely different perspective on evil and suffering: We are no longer "good" people

---

15. John 3:17.

who suffer "bad" things; we are "bad" people who enjoy so many "good" things. So never turn misery into self-pity. As Jesus once said, "No one is good but God alone."[16] And besides, who is to say suffering is all bad, or bad forever? We said earlier that God is certainly able to use suffering for a better purpose, for something good. There is something therapeutic about suffering—it has the potential to redeem us, transform us, and transfigure us.

Again, do not think I am glorifying suffering here, but God may use evil and suffering, which befell us or were caused by us, for a better good, for a better purpose—some might even say to discipline us, to purify us, to correct us, to test, and to teach us. In everything that happens, we can discern God's "hand"—not a hand that *causes* all the good and bad things that we read about in the newspapers, but a hand that *holds* all these things together by saving them for a better purpose and destination. It is in such a "hand" that God holds the future.

And then there is also a much more important goal for suffering: It can be used for our redemption. Redemption is actually a culmination point in the Bible. We find it already in the Jewish Scriptures when some take upon themselves the sins and burdens of others so that all will be free of the consequences of sin. It is sometimes called vicarious atonement.

We find this belief already in the famous story about Abraham's plea for Sodom.[17] He asks God, "Will you indeed destroy the righteous with the wicked?" Then Abraham starts negotiating: Suppose there are fifty righteous within the city, would you then destroy the place and not spare it for the fifty righteous who are in it? Next Abraham whittles this farther down: What about forty-five? What about forty? What about thirty? Suppose twenty are found there? And at last he gets God down to ten: "For the sake of ten I will not destroy it."

The previous story even has a sequel, counting down from ten to *one*, when God says to the prophet Jeremiah: "Run back and forth through the streets of Jerusalem [...] Search her squares to see if you can find a man, one who does justice and seeks the truth; that I may pardon her."[18] Ultimately, that one man will be found in Jesus of Nazareth, the Son of Man. It was

16. Mark 10: 18. Luke 18:19.

17. Genesis 18:16-33.

18. Jeremiah 5:1.

the High Priest Caiaphas who spoke to the Sanhedrin the prophetic words: "One man should die for the people."[19]

There is something very peculiar going on here: We tend to ask God not to treat the just the way the unjust deserve to be treated, but instead God decides to treat the unjust in the same way as he treats the just, as if they were just too. Believe it or not, God in fact somehow pardons the unjust majority because of the just minority. That was also the "mission" Jesus Christ had—through him humanity can be pardoned. Jesus gave us our dignity back, but at the cost of his own. Each time we ask God "Why me?" we should listen carefully to hear Jesus whisper in response "Why me?"

Although the Book of Job—about this honest and just man who suffered so many afflictions—plays a "crucial" role in the Jewish Scriptures, Christianity is really the only religion that places the Cross center stage, in the midst of the black holes in the universe and in the midst of the black holes in our personal lives. But is the Cross also at center stage in God's providence, you might ask. Yes, it is! When Abraham said to his son, Isaac, "God will provide himself the lamb for a burnt offering,"[20] he was referring to God's providence.

Christians see the story of Abraham and Isaac as a profound allegory for the sacrifice of Jesus on the Cross. Like Isaac, Jesus was a father's only beloved son. Like Isaac, Jesus carried uphill the wood for his own sacrifice. Even Abraham's words proved prophetic. Since there was no punctuation in the Hebrew original, verse 8 could be read as follows: "God will provide Himself, the Lamb, for a burnt offering"—the Lamb being Jesus Christ, God himself. So the Cross is not mere darkness, since it is also part of God's providence. The ancient Church used to sing "God reigns from the wood of the Cross." In other words, suffering is not meant to make us bitter but better.

## A Universe of Winners and Losers?

We have at least one more very delicate issue left that has caused much turmoil and confusion in the history of Christianity.[21] If God is all-knowing—

19. John 11:50.

20. Genesis 22:8.

21. A good discussion can be found in: James Akin, "A Tiptoe through Tulip," *The Rock*, Volume 4, 8 (September 1993).

as he is, so we found out—then he must have *foreseen* our future; he must have foreseen that some will end up in Heaven and others in Hell. If God knows everything, he must also know who will be saved and who will be damned. Does this not sound like a world of "born winners and born losers"?

This issue is usually referred to as *predestination*. It is very well known in some Christian circles, especially of the Calvinist tradition, but it is also part of the Catholic faith. Although Christian churches differ in their interpretation, all believe in some form of predestination, because the Bible uses the term. The apostle Paul says, "these whom he predestined, he also called,"[22] and "he predestined us for adoption to himself through Jesus the Messiah, according to the pleasure of his will."[23] Predestination is definitely part of God's providence in Christianity.

But does this not suggest that God has his favorites—those predestined to salvation versus those predestined to damnation? Are some of us predetermined to be winners and others losers? Let us find out. Thomas Aquinas put it this way: "The reason for the predestination of some, and reprobation of others, must be sought for in the goodness of God. [...] God wills to manifest his goodness in men; in respect to those whom he predestines, by means of his *mercy*, as sparing them; and in respect of others, whom he reprobates, by means of his *justice*, in punishing them. This is the reason why God elects some and rejects others."[24]

These are surely heavy words. Do they mean there is an eternal, predetermined division between two kinds of people? Not so in Aquinas' thinking! God first intends the purpose of the universe, so it might manifest his goodness as far as is possible, including his mercy as well as his justice. Therefore, he intends that there should be salvation as well as damnation, so that his mercy and justice might be more perfectly manifested. God first decides the end and then decides the adequate means to that end, namely *predestination* and *reprobation*, the ultimate end being the same for either, namely the manifestation of his divine goodness, which implies mercy as well as justice—neither one without the other.

---

22. Letter to the Romans 8:30.

23. Letter to the Ephesians 1:5.

24. Aquinas, *Summa Theologica* 1, 23, 5.

Apparently, in the end, it does matter what kind of person you and I have been. As we saw earlier, God's willing that I decide as I do does not make my decision God's. God's willing does not take away from my own free will, or the actions founded upon them; they remain my own. But make no mistake, there is an important consequence as well: Any sin they involve also remains mine. That is where sin comes in. Sin is the dark side of human freedom. Having free will means having a nature that is incomplete, in the sense that what we *are* never fully determines what we will *do*. By establishing an identity for ourselves, we may very well find ourselves in rebellion against God, simply because as free beings we might assume that we can establish our destiny fully on our own in a way that escapes providence.

On the other hand, to be "friends" with God requires, on our part, a responsible decision to accept the offer of friendship he presents to us. But a responsible choice in God's favor requires that we understand the alternative too—which is to be at enmity with him, cutting us off from our Creator. As Creator, God is fully involved in all our acts, including those in which we sin, for they can occur only through his will. Yet, God incurs no blame for them, as they are our acts, not his. When we do something wrong, it would be nice and convenient if we could blame God for it—but alas, we can only blame ourselves.

Because we are free human beings, we will be held accountable for our choices in life—and that is where Hell or eternal damnation comes in. God does not cast anyone into Hell against their will. No one wants Hell to exist. No one wants evil to exist. But Hell is just evil eternalized. No one can say, "See no evil." If there is evil and if there is eternity, then there can be Hell—it is a matter of our own free choice. C. S. Lewis called Hell "the greatest monument to human freedom."[25] He also says, "There are only two kinds of people—those who say to God, 'Thy will be done,' or those to whom God says in the end, 'Thy will be done.'"[26]

Indeed, it is hard to see how a loving God could do anything but honor our choice in the matter. If we do end up in Hell, we are not "rejects" rejected by God from the start, but we have rejected ourselves. If you wonder how God could ever create human beings that he knew would end up

25. C. S. Lewis, *The Great Divorce* (San Francisco: Harper, 2001).

26. C. S. Lewis, *Mere Christianity* (San Francisco: Harper, 2001).

in damnation, you should ask yourself if God would have been more loving had he *not* created them. Even in the lives of the lost, God is lovingly involved, as they are full participants in God's creation and providence, notwithstanding the fact that they decided to ruin their own lives by securing their tragic destiny and rejecting his providence.

Could God not have done this differently? I do not believe so. It would really be meaningless to suppose that God would have shown greater love toward the ones who choose to be lost by omitting them from creation—for what is not there cannot be loved. Equally, it is meaningless to think the lost would be better off had they not existed, for what does not exist is neither well nor bad off. And it is as good for the lost as it is for the saved to have the opportunity for salvation, and to have a choice as to whether to accept it. But what is *not* good for the lost is the fact that they decided to reject God. Yet that is fully their decision, and its consequences are fully earned. But we should never accuse God of rejecting those people! Augustine would say, "God created us without us: but he did not will to save us without us."[27]

So who will make it to Heaven? Is it only the happy few? No, God wants *everyone* to go to Heaven, but everyone also has the free will to decline or accept his offer.[28] Theologians such as John Calvin, however, keep maintaining that our capability of choosing would be a direct violation of God's sovereignty. They claim that if human beings are able to frustrate God's desire to bring everyone to salvation, then God's will would not be all-powerful. Human beings would be more powerful than God, as they have the power to frustrate God's plan and power. Therefore, if certain people are not saved, then it is supposedly not because they *chose* not to be saved, but because God, in his sovereignty, *willed* or predestined that they would not be saved.

If that were really true, then we would have to believe that God desires only the salvation of the chosen ones, the elect, the "happy few." Indeed, John Calvin accepts this consequence and says that whom God selects,

---

27. Augustine, *Sermo* 169, 11, 13: PL 38, 923.

28. Often Matthew 22:14 is quoted here: "Many (πολλοί) are called but few (ολιγοι) are chosen." First of all, "many" in English is restrictive in the sense of "many, but not all," whereas in Greek it is inclusive, equivalent to "practically everyone." Second, the "chosen" are those who choose to accept.

he saves, but whom God rejects, he damns.[29] While studying for a defense of Calvinism, the Dutch theologian Jacob Arminius became convinced, around 1600, that he was defending an illogical position and sought to modify Calvinism so that God might not be viewed the author of sin, nor man an automaton in the hands of God.

I would say Arminius and his followers, the Remonstrants and Arminians, are right. If we let God alone be responsible for everything, not only do we make God completely responsible for the salvation of the *saved* ones, the "elect," but also for the damnation of the *lost* ones, the "reprobate," even to the point of directly willing their sins. If so, the elect would be saved with no merit of their own, and the reprobate would be damned for no fault of their own. This may "save" God's Sovereignty, understood in a very restrained sense, but surely not our freedom and not God's goodness, mercy, and justice.

This stands in sharp contrast to what the Catholic tradition holds. Catholics would argue: Being all powerful, God—and God alone—has the power to limit the sphere in which his will is effective. God's power is at a level completely different from our human powers. By his own will, God has chosen to limit the realm of his own power, creating beings that have a free will that can frustrate his own will. In one of his encyclicals, Pope John Paul II writes: "Since salvation is offered to all, it must be made concretely available to all."[30] The Catholic Church also teaches that "God predestines no one to go to Hell."[31] Hence, Catholics argue that salvation is made available to all and willed for all, even for those who are not explicitly part of their Church; all are thus predestined for salvation, for the Church is not an organization of saints but a hospital for sinners. It is not material prosperity for all that the Gospel proclaims but rather spiritual salvation for all.

Not only is God's providence *all-loving* but also *loving-all.* Although election and salvation depend on God's infallible fore-knowledge, we just have no way yet of knowing to which category you and I will belong at the end. Nevertheless, not one of us can ultimately frustrate God's overall plan, and this is the truth Calvinists want to protect as much so as Catholics.

---

29. John Calvin, *Institutes of the Christian Religion*, Book III, 21, 7.

30. *Redemptoris Missio,* 10.1.

31. Catechism of the Catholic Church, 1037.

Augustine once said, "Just as one can find that which is not Catholic in the Catholic Church [...], one can also find something that may be Catholic outside of the Catholic Church."[32]

Let us make one more time this vital, important distinction between God's providential will and God's permissive will. Aquinas would say that, because of God's *permissive* will, God permits evil events to happen so as to pull a greater good out of them. What precisely that "good" is, in the long run, we can never know for sure. In contrast, God's *providential* will is God's desired activity in the world. The ultimate goal of God's providential will is the salvation of all humanity. But God does permit the fact that some throw their chance of salvation away at their own decision. We are all *predestined* to salvation, but this does not mean we are all *predetermined* to salvation.

So do not confuse pre-destination with pre-determination, for they are in fact quite different concepts. In the words of Augustine, "Predestination is nothing else than the fore-knowledge and fore-ordaining of those gracious gifts which make certain the salvation of all who are saved."[33] God's fore-knowledge cannot force our free will, for the simple reason that it is basically nothing else than the eternal vision of the future of human decisions, as we discussed earlier—which does not deny or exclude free will at all. God knows with certainty his elect, but we do not. God fore-sees, as in an eternal pre-view, the free activity of everyone precisely as that individual is willing to shape it. So pre-destination is not pre-determination of the human will. No one is predetermined to get saved or lost; everyone is predestined to salvation, but not everyone accepts it.

Yet, this confusion has also been the source of another misconception in Christian history. If pre-destination were actually a matter of pre-determination—which it is not, as we found out—then we would have no reason to better our lives, for nothing we do would change the outcome anyway. It is like watching the replay of a game on TV. Does it matter whether the players tried to win? Of course it does, but when you watch a replay, it has already been decided who the winners are. God knows the outcome of the "game" we are in, but we, who are still in the game, have no way of knowing what the outcome will be. "Strive to enter...," says the

32. Augustine, *On Baptism, Against the Donatists*, PL 43 , VII, 39, 77.

33. Augustine, *Perseverance*, 14:35.

Bible. We are all invited to win. It is clear that we know ahead of time there are winners, but we do not know yet who they will be. We all have to "run" to see who will make it to the finish line.

If pre-destination were just a matter of pre-determination, another erroneous consequence would be that grace could only be given to the "pre-chosen" ones, forcing them into a state of permanent grace that presumably can never leave them until they die. Grace would be irresistible—a gift that you cannot refuse (but a gift you cannot refuse is not really a gift, I would counter). It is exactly because of this line of thinking that some Christians have come to believe they can no longer fall away, since they are saved forever. They sometimes call themselves "born-again," and believe that a profession of faith is all that is necessary to procure salvation. But there is something missing here, just as there is something missing when you profess your marriage vows and then think you need no longer express your love through a lifetime of loving actions.

Again, pre-destination is *not* pre-determination: If Adam and Eve, who possessed grace and a perfectly intact nature, could freely sin, how much more so is it possible to sin for all their descendants—including those so-called born-again Christians who possess grace but also a wounded nature and a darkened intellect. If Adam and Eve could fall from grace, surely all of us can fall from grace as well. Hence, do not confuse "the elect" with the "born-again." The elect are those who persevere to the end, but not necessarily those who were once born again and started out the right way. Reality is that born-again Christians can and (sadly) do fall away sometimes.

Faith is not a once and for all event; it must be preserved, nourished, and cherished—which is a task of a lifetime. Through grace, we know we desperately *need* mercy, for we do not have it on our own and cannot claim it as our own. Grace and law, heart and head, faith and reason must be brought together in the service of God.

If there is a division between those who are elect and those who are not—and there is—then it is of a very different nature. The apostle Paul would characterize it thus: "God chose what is foolish in the world to shame the wise, God chose what is weak in the world to shame the strong" (1 Cor. 1:27).

# 7. Conclusion

Is there anyone out there providing for us? Does the Universe have a destination? Those are the main questions we tried to answer in this book. In our pursuit of the Great Unknown, we have reached a stage where I am happy to answer these questions with a definite yes—and I hope you do too. But let me be clear about it: The Great Unknown is not someone out there in space, but someone "outside" and "beyond" space—and yet "inside" each one of us—someone who "holds" the universe and each one of us "in his hands" and keeps them in existence.

We are indeed living in a purpose-driven universe, driven not only by purposes in the minds of its inhabitants but also by purposes in the Mind of its Creator. Obviously, the *destiny* of the universe goes far beyond the *fate* of the universe. It is placed under the guidance of *providence*, assuring us that everything unfolds in accordance with God's plan—and that is what gives religion its strong anchor. Even someone like the philosopher Ludwig Wittgenstein could finally say, "Religion is, as it were, the calm bottom of the sea at its deepest point, which remains calm however high the waves on the surface may be."[1]

Since Copernicus, we may have lost our central position in the universe and since Darwin we may have lost our unique position in the animal world, but that does not mean we have also lost our personal significance in the universe—in spite of the fact that someone like the late astrophysicist Carl Sagan kept hammering that "we live on an insignificant planet of a humdrum star lost in a galaxy tucked away in some forgotten corner of a universe in which there are far more galaxies than people."[2] Forgotten? Forgotten by whom, I would ask. Sagan was still thinking in the line of the Supreme Court Judge Oliver Wendell Holmes Jr. (1902-1932) who once wrote, "I see no reason for attributing to man a significance different in

---

1. Ludwig Wittgenstein, *Culture and Value*, translated by Peter Winch, (Oxford: Basil Blackwell, 1980), 53e.
2. Carl Sagan, *Cosmos, One Voice in the Cosmic Fugue*, (New York: Random House, 2002), 31.

kind from that which belongs to a baboon or a grain of sand."[3] No significance, really? No significance to whom? Do we really have to believe these slogans and go back again to where we started in this book? I hope not.

Let us not forget that each one of us is an "operating center" where rational and moral decisions are being made. Everybody is somebody; each one of us is not just an evolutionary nobody but rather a significant somebody. We are the only ones on earth who can declare ourselves significant—or insignificant, for that matter. We can even downgrade "the" human mind while touting our own minds. The Harvard astronomer and science historian Owen Gingerich puts it well: "We human beings are the most extraordinary creatures we know about, and part of our glory is that we can imagine we are not the most remarkable creatures in the entire universe."[4]

Even if we call ourselves insignificant beings in this universe, we should not forget that only *we* are able to do so—because there is an immense universe inside each one of us. We can even say about ourselves that we are "only human"—thus comparing ourselves, not with something "below" us (such as a cat, a dog, or an ape), but with something "above" us and transcending us. I cannot transcend myself on my own, of course, but because I myself was made in the image of God, I perceive more than myself whenever I perceive myself completely—and in this process, I may become more like my Maker.

What gives us genuine significance in life is the fact that God, who brings everything into being, called us *personally* into being, because we are made in his image, and he endowed us with a rational and moral mind. That is why each one of us is a significant somebody; each one of us is the result of a thought of an all-loving God who is also a loving-all God. He calls each one by name, but it is up to us whether we answer the call. Although we came into this world alone and will leave this world alone, we are not really alone, for we came here from our Maker and will eventually return to our Maker, when he calls us back by our name. In the meantime

---

3. Letter to Frederick Pollock, August 30, 1929, in Richard A. Posner, ed., The Essential Holmes: Selections from the Letters, Speeches, Judicial Opinions, and Other Writings of Oliver Wendell Holmes, Jr. (University of Chicago Press, 1992), 108.

4. Owen Gingerich, *God's Universe* (Cambridge, MA: Belknap Press of Harvard University Press, 2006), 1st lecture.

he envelops us as an all-present God and embraces us as an all-loving God. As Pope Benedict XVI put it in his first homily as Pontiff, "We are not some casual and meaningless product of evolution. Each of us is the result of a thought of God." Therefore, life's journey stretches from before the womb to after the tomb.

That is also what I discovered more and more clearly in the course of my own life, and I hope you did as well, perhaps in reading this book. Many others have experienced something similar. Let me mention only one of them more in particular, a witness beyond suspicion—the French philosopher and renowned atheist Jean-Paul Sartre.[5] Toward the end of his life, by then blind, in poor health, but still in full possession of his faculties, the man whom most people know as an uncompromising atheist had a profound conversion. In the early spring of 1980, he shared much of his time with an ex-Maoist, Benny Lévy (writing under the pseudonym Pierre Victor), and the two had a dialogue in the ultra-leftist *Le Nouvel Observateur*. It is sufficient to quote a single sentence from what Sartre said during this dialogue: "I do not feel that I am the product of chance, a speck of dust in the universe, but someone who was expected, prepared, prefigured. In short, a being that only a Creator could put here; and this idea of a creating hand refers to God."[6] Imagine, Sartre's physical universe had become a meaningful universe! He was finally able to say, "He leads the way, for God is His Name."

Doesn't this sound like a profession of faith—actually much to the consternation of his life-long girlfriend Simone De Beauvoir, who spoke of the "senile act of a turncoat"? It seems to me that, at last, this blind, old man, who had once written, "Both your life and your death are merely nothing,"[7] had been cured from his "mental myopia." He finally had seen the light at the end of the tunnel! At last, in his subconscious pursuit of the Great Unknown, Sartre was able to see the destiny of our universe not as a fate but a plan born in the Mind of the Great Unknown.

---

5. The late Catholic philosopher Joseph M. Bochenski described Sartre as "probably the most intelligent and astute atheist that history has ever witnessed." In *The Road to Understanding*, (North Andover, MA: Genesis Publishing, 1996), 119.

6. Quoted in *The National Review*, November 6, 1982.

7. Jean-Paul Sartre, *The Condemned of Altona*. Trans. Sylvia and George Leeson. (New York: Alfred Knopf, 1961), 171.

I want to join Sartre in saying that we are all heading for a great destination. God's providence is there for us—not in a dictatorial or autocratic way, but in a merciful and compassionate way. Having said that, I would also like to join Aquinas in the confession he made at his death bed: "Thee have I preached; Thee have I taught. Never have I said anything against Thee: If anything was not well said, that is to be attributed to my ignorance."[8]

8. Quoted by Richard J. Foster and Gayle D. Beebe, *Longing for God.* (Downers Grove, IL: IVP Books, 2009), 128-129.

## Suggestions for Further Reading

Ayala, Francisco J. *Darwin's Gift to Science and Religion.* Washington, DC: Joseph Henry Press, 2007.

Ayala, Francisco J. *Am I a Monkey? Six Big Questions about Evolution.* The Johns Hopkins University Press, 2010.

Bancewicz, Ruth (ed.). *Test of Faith: Spiritual Journeys with Scientists.* Eugene, OR: Wipf & Stock, 2010.

Barbour, Ian G. Religion and Science: Historical and Contemporary Issues. San Francisco: HarperCollins, 1997.

Barr, Stephen M. *Modern Physics and Ancient Faith.* South Bend, IN: University of Notre Dame Press, 2006.

Bochenski, Joseph M. *The Road to Understanding: More than Dreamt of in Your Philosophy.* North Andover, MA: Genesis Publishing Company, 1996.

Feser, Edward. *Aquinas: A Beginner's Guide.* Oxford: Oneworld Publications, 2009.

Gingerich, Owen. *God's Universe.* Cambridge, MA: Belknap Press of Harvard University Press, 2006.

Hannam, James. God's Philosophers: How the Medieval World Laid the Foundations of Modern Science. London: Icon Books, 2009.

Jaki, Stanley L. *The Savior of Science.* Grand Rapids, MI: Eerdmans, 2000.

Lewis, C. S. *Mere Christianity.* New York: Harper Collins, 2001.

Miller, Kenneth R. Finding Darwin's God: A Scientist's Search for Common Ground between God and Evolution. New York: Harper Collins, 1999.

Van den Beukel, Anthony, *The Physicists and God: The New Priests of Religion.* North Andover, MA: Genesis Publishing Company, 1995.

# Index

## D



## E



## F



## G

## U



## V



## W

# About the Author

Dr. Gerard M. Verschuuren is a human geneticist who also earned a doctorate in the philosophy of science. He studied and worked at universities in Europe and the United States. Currently he is semi-retired and spends most of his time as a writer, speaker, and consultant on the interface of science and religion, creation and evolution, faith and reason.

His most recent books are:

- *Darwin's Philosophical Legacy—The Good and the Not-So-Good* (Lanham, MD: Lexington Books, 2012).
- *God and Evolution?—Science Meets Faith* (Boston, MA: Pauline Books, 2012).
- *Of All That Is, Seen and Unseen—Life-Saving Answers to Life-Size Questions* (Goleta, CA: Queenship Publishing, 2012).
- *What Makes You Tick?—A New Paradigm for Neuroscience* (Antioch, CA: Solas Press, 2012).
- *Life's Journey—A Guide from Womb to Tomb* (upcoming).
- *It's All in the Genes!—Really?* (upcoming).

For more info: http://en.wikipedia.org/wiki/Gerard_Verschuuren.
He can be contacted at www.where-do-we-come-from.com.